Microsoft
Windows 7
A Beginners Guide

By WR Mills

AuthorHouse™
1663 Liberty Drive
Bloomington, IN 47403
www.authorhouse.com
Phone: 1-800-839-8640

First published by AuthorHouse 12/21/2009

ISBN: 978-1-4490-4732-0 (e)
ISBN: 978-1-4490-4731-3 (sc)

Library of Congress Control Number: 2009913336

Printed in the United States of America
Bloomington, Indiana

This book is printed on acid-free paper.

Microsoft
Windows 7
A Beginners Guide

A training guide for Windows 7

About the Author

Bill has a background in electronics and technology. He started writing software in 1982 and has expanded his programming skills to include C, C++, and Visual Basic. Bill also designs web sites. He designed a computer based telephone system for the hotel/motel market and wrote the entire operating system himself.

In 2007 he started teaching computer training classes and seems to have a knack for explaining things in a simple way that the average user can understand.

Bill is self-employed and lives in Branson Missouri. He has three children, two sons and a daughter.

Preface

In 2007 I started teaching computer training classes. I was shocked at how much trouble the students had trying to understand the textbooks. I spent all of my time explaining what the textbook was trying to get across to the reader. It wasn't until I started getting ready for teaching the Microsoft Office 2007 series of classes that I finally gave up and started writing the textbooks myself.

These books are easy to understand and have step by step, easy to follow, directions. These books are not designed for the computer geek; they are designed for the normal everyday user.

It seems I have a knack for explaining things in a simple way that the average user can understand. I hope this book will be of help to you.

William R. Mills

Foreword

Dear Bill, I wanted to write a note of appreciation to you for your books: Microsoft Office Word 2007, Microsoft Office Excel 2007, and Microsoft Office PowerPoint 2007. I've used them all and found each one to be easy to read and very user friendly. If anyone needs to learn one of the 3 programs, but is even a little intimated, I strongly suggest they try one of your books. It's almost as good as taking a class with you as the instructor. If I didn't understand a step, I just went back to the pervious step and tried it again --- and it always worked!! There's just enough humor in the text to keep the reading interesting; never dull, but fun and light. Just what a beginner needs. Again, Bill, I thank you for creating these books that make learning something I needed to learn fun and easy. Sincerely, Cyndy O

Important Notice:

There will be times during this book that you be asked to open a specific file for the lesson. These files can be downloaded from the EZ 2 Understand Computer Books web site.

Open your internet browser (probably Internet Explorer) and go to www.ez2understandcomputerbooks.com. Click your mouse on the Lesson Files link toward the top. This will take you to the page where you can download the files needed. There are directions on the page to help you with the download. They are repeated below for your convenience.

To download the files follow the following steps:

1)   Right-click your mouse on the file(s) you want. These are zipped files and contain the lessons that you will need for each book.

2)   Select the "Save Target As" choice. Make sure the download is pointed to a place on your hard drive where you can find it, such as My Documents.

3)   Click the Save button

4)   The files are zipped files and will need to be extracted to access the contents. To extract the files right-click on the file and choose "Extract All".

Table of Contents

Chapter One   The Basics

What is Windows 7?

Windows 7 is the newest and most exciting operating system from Microsoft. An operating system is a sophisticated software program that controls not only the way you work on your computer, it also controls just about everything that is connected to your computer. If my memory serves me correctly, Windows got its name from all of the little windows the operating system puts on your monitor. I mention this because you can put several of these "little windows" on your monitor screen at one time and jump back and forth between the different windows. This will allow you to work with several different programs at one time. Windows 7 will control each of these programs and keep everything from getting mixed up as you work with them. That in itself says a lot.

Perhaps the best way to describe Windows 7 is to say that it is an improved version of Vista (with the fixes it needed) and a few of the Apple features thrown in. I, for one, was eagerly waiting for Windows 7 and couldn't wait to start playing with it. So far I have been having a lot of fun. I think that you will too.

Chapter one will cover such things as starting Windows 7 and logging in. We will also cover the things you need to know to have your computer look and act the way you want. You may be surprised to know that Windows 7 is not really difficult to use. In fact I think that what will surprise you is that is very easy to use.

So, take a deep breath and let's get started.

Lesson 1 – 1   Starting Windows 7

Listed below are the "proper steps" to turning on your computer.

1) Take a deep breath

2) Exhale

3) Push the on/off button

Okay, go ahead and try it

After you have performed the above steps, Windows 7 will start automatically. In a few moments, give or take, Windows 7 will jump onto your screen. It is not a fast jump, but it is considerably faster than some of the previous versions.

If you have a password on your computer login name, you will see the Login screen, sometimes called a Welcome Screen, and you will need to enter your password and then press the Enter key on the keyboard. If a password is not required, you will see your desktop on the screen. If there is more than one user on your computer, you will still go to the Login screen and you will have to click on your user name, but you won't have to put in a password, unless you have required a password for your login name.

As you enter your password, if it is required, a series of ****** will be displayed on the screen. This is to prevent anyone from seeing your password as you type. This is a safety feature for your protection. So what if you forget your password? Should we throw our computer away and buy another one? Microsoft wouldn't do that to you. If you enter an incorrect password, the next screen will have the helpful hint that you provided when you set the password. It will be directly underneath the textbox where you type your password. You should know that whoever tries to login to your user account will be able to see this helpful hint. So don't put something like "The password is Bill". Put something that will remind you of what the password is.

The desktop comes in many colors and has many pictures available so that your screen will look more attractive when you look at it. Since this will not affect the way your computer works, we won't deal with changing the picture on the screen until Chapter two. Figure 1-1 shows an example of the login screen.

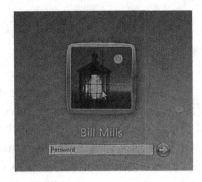

Figure 1-1

Lesson 1 – 2   The Windows 7 Screen

When you start Microsoft Windows 7, you will notice that the screen looks similar to the Windows Vista screen. If you have never used the Microsoft Windows Vista this could look confusing and overwhelming. The opening screen is called the Desktop. This provides a background for all of the other windows that will appear on the monitor. Figure 1-2 shows a typical desktop.

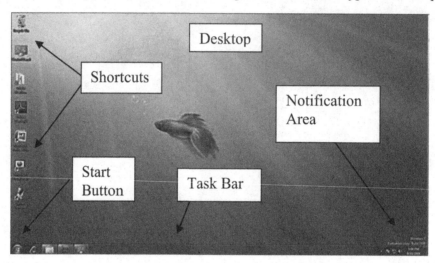

Figure 1-2

As I said the entire screen is called the Desktop. The color on your desktop and the picture shown on your screen may be different than the one shown in the figure.

The icons shown on the left side are shortcuts. These allow you to quickly access some of the programs that you will be using a lot of the time. The shortcuts on your monitor may be different than the ones on my screen. But fret not these can be changed and we will cover just how to do that in Chapter two.

The Start button is the round circle in the lower left corner of the screen with the Windows Logo on it. In past versions (excluding Vista) this was a square button that had the word Start on it. This is where, as the name implies, you start the different programs. We will deal with this more in the next lesson.

The Task bar is the bar going across the bottom of the screen. This holds the Start button on one end and the Notification area on the other. In the center are buttons for the programs that are currently being used. You can also have some programs permanently showing on the Task Bar for quick access, The Task Bar will be discussed in Lesson 1-7.

Some of the things on the Desktop should look familiar to you if you have ever used a computer with Windows on it. The icons for the shortcuts are the same as the earlier versions of Windows. The Task bar was also used in the earlier versions of Windows. The Notification area, where the clock is located, was used in earlier versions of Windows.

The earlier versions of Windows did have one other thing on the Task Bar that is not in the Windows 7 version. This is the Quick Launch Bar. This has been combined as part of the Task Bar.

By now you are tired of just reading, so let's actually do something.

Lesson 1 – 3   Using the Mouse
Pointing, Clicking, and Double-clicking

The mouse is a small device that will fit into the palm of your hand and lets you point at, select, and move objects on your computer screen. The mouse is linked to a pointer on your computer screen, when you move the mouse, the pointer moves on your computer screen. This lesson will show you how to perform the most basic mouse actions.

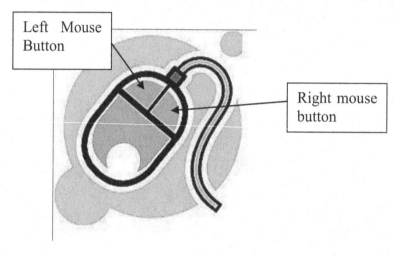

Figure 1-3

Rest your hand on top of the mouse, then move the mouse and watch as the arrow moves across the screen.

The arrow (also called the cursor or pointer) you see on the screen follows the mouse movement as you move your mouse across the desk or mouse pad.

One of the most basic things you can do with a mouse is pointing. To point to something (an object), simply move the mouse until the pointer is touching the object.

Move the mouse pointer until the tip of it is over the Start Button. Leave the mouse pointer there for a few seconds.

A screen tip will appear after a few seconds that says "Start". To use the start button you need to click it with the mouse. "Clicking" means pressing and then releasing the mouse button. The mouse makes a clicking noise when you press and release either mouse button, hence the name clicking.

6

Move the pointer over the Start Button and press and release (click) the left mouse button once.

When you click the Start Button the Windows 7 Start Menu pops up. You can close the Windows 7 Start Menu without choosing anything by clicking anywhere outside of the Start Menu.

Note: The mouse has two buttons, a left button and a right button. Normally you will use the left button. In these lessons you can assume that when it is stated to click on something, I mean you are to use the **left mouse button**. The right mouse button has its own purpose, and we will discuss that later.

Move the pointer anywhere outside of the Start Menu and click the left mouse button.

Now that you have used the mouse and feel a little more comfortable clicking it, we will move on to something a little harder: double-clicking. Double-clicking is exactly as it sounds, clicking the mouse button twice in rapid succession. You will usually use double-clicking to open an object, file, or folder.

Move your mouse pointer over to the Recycle Bin and double-click it with the left mouse button.

The Recycle Bin icon is shown in figure 1-4.

Figure 1-4

The Recycle Bin will open and reveal its contents as shown in figure 1-5.

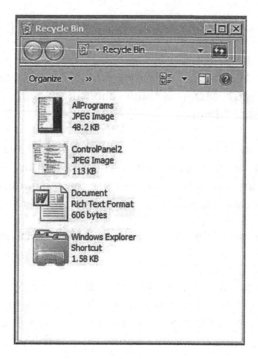

Figure 1-5

The contents of your Recycle bin will probably not look like the figure. The figure shows the contents of the Recycle Bin on my computer. Your Recycle Bin may not even have anything in it. The Recycle Bin is where the things that you delete from your computer end up.

Note: A lot of people have trouble double-clicking the first few times they try. If you are having trouble, it is probably because you are not holding the mouse steady or you are not double-clicking fast enough. If you click the mouse button too hard, the mouse will move slightly and Windows will think that you have made two single-clicks and not one double-click.

Close the Recycle Bin by clicking its Close Button, the small box in the upper right corner with the X in it.

The following table tells you what you can click and double-click.

When you want to:	Do this:
Select something (an object)	Single-click the mouse
Open a menu	Single-click the mouse
Press a button on a toolbar	Single-click the mouse
Move to a new area in a dialog box	Single-click the mouse
Open a file	Double-click the mouse
Open a folder	Double-click the mouse

Table 1-1

In the next lesson we will practice using the mouse.

Lesson 1 – 4   Using the Mouse
Clicking and Dragging

You can move items around on your computer screen by clicking and dragging them with the mouse. There are three steps involved with clicking and dragging, they are as follows:

Move the mouse pointer over the object you want to move, then click and hold down the left mouse button.

While you are still holding the left mouse button down, move the mouse until the pointer is over the place you want to put the object.

Release the left mouse button.

The rest of this lesson may seem silly to you, but you need to master clicking and dragging. Therefore your assignment for this lesson is to play a game, a game of Solitaire. Some people believe that Microsoft still includes this game as a learning tool for clicking and dragging. Either way it is fun and good practice. The following steps will show you how to start the game.

Click the Start Button in the bottom left corner of the screen.

On the right side of the Start Menu, click the word Games

Use the scroll bar on the right side of the menu box that came onto the screen, to go down to where you see Solitaire.

Double-click the card and word Solitaire.

Play a game of Solitaire and practice clicking and dragging the cards.

If you don't know how to play solitaire, here is a quick recap of the game. You move lower numbered card on top of higher numbered card (a queen goes on top of a king etc.), colors must alternate (a red card on top of a black card), a group of cards can be moved if the very bottom card can be played on top of the card where you are moving it to. Aces go up to the top and the next card in the sequence may be placed on the Ace (The Ace of Spades is placed at the top, the two of Spades is placed on the Ace, the three of Spades is placed on top of the two of Spades, etc.). To move a card (or group of cards) simply click on the card (keep holding the mouse button down) and drag it to the new location and release the mouse button. If you drag it to an invalid place it will whisk back to where it started.

Click the close button when you are finished playing.

There are other things that you can click and drag besides cards. We will cover these later, but in the mean time table 1-2 lists some of examples of when you can use click and drag.

Table 1-2 – Things you can click and drag

What you can do:	How you do it:
If you want to move a window to a new location on the screen	You drag a window by its Title Bar and release the mouse button to drop it to its new location on the screen
If you want to move a file to a new folder	You drag the file and release the mouse button when it gets to the desired folder
If you want to change the size of a window	You drag the corners or borders of a window to make it larger or smaller
If you want to scroll down a window to see something located off-screen	You drag the scroll box up or down the scroll bar and stop when it gets to the desired location
If you want to move just about anything on your computer screen	Point to the object, click and hold down the left mouse button, drag the object to a new place, and then release the mouse button

Table 1-2

Lesson 1 – 5   Using the Mouse

Right-clicking

You already know that the left mouse button is the primary mouse button that you will use. It is the one used for clicking and double-clicking. The obvious question is what is the right mouse button used for? Whenever you right-click something it brings up a shortcut menu to the screen that lists everything you can do to the object. If you are unsure about what you can do with an object, point to it and click the right mouse button. A shortcut menu will appear on the screen with a list of commands that are related to the object or area you right-clicked.

Move the pointer over the Recycle Bin and click the right mouse button.

A shortcut menu will appear with a list of the commands related to the Recycle Bin as shown in figure 1-6.

Figure 1-6

If it is available, point to and click the Empty Recycle Bin **option on the shortcut menu.**

You will use the left mouse button to select a menu item even if you used the right mouse button to open the shortcut menu. A dialog box will appear asking if you are sure you want to delete the contents of the Recycle Bin. Since we don't know what is in the Recycle Bin, let's play it safe and not delete everything.

Click No with the left mouse button.

Now, let's suppose that you have realized that the clock in the bottom right corner of your computer screen did not automatically reset for daylight savings time, because the government decided to change everything again. How are you going to fix that problem? You can display the clock's properties by right-clicking on the clock.

Move the pointer over the clock (located on the far right side of the Task Bar) and click the right mouse button.

Another shortcut menu appears, with commands related to the Notification area, as shown in figure 1-7.

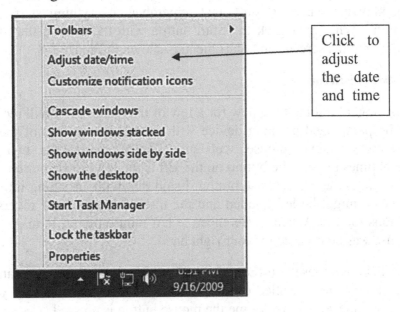

Figure 1-7

You will notice that one of the commands is Adjust Date/Time. If you wanted to change the date or the time, you would select this option from the shortcut menu. At this time, let's NOT change anything. Now all we need to do is close the shortcut menu without selecting anything.

Move the mouse anywhere outside the shortcut menu and click the left mouse button.

That is all there is to it, you can now point, select, click, double-click, click and drag, and right click the mouse. Now you can take five, get a cold drink, and pat yourself on the back for a moment, then we will move on to the next lesson.

13

Lesson 1 – 6   The Start Button & Menu

The Start button is the main way you will open programs. If you find yourself not knowing where to start, just remember "everything starts with the Start button". When you click the Start button with the mouse, the Start Menu will jump onto the screen. The Start menu is shown in Figure 1-8.

Click the Start Button

This might be something new for a few of the readers so I will recap. The mouse is the palm sized pointing device with a cord coming out of the back of it. It resembles a small mouse, well sort of. The mouse has two buttons on it (sometimes three). The button on the left is the left mouse button and the one on the right is the right mouse button. I said that without taking into consideration that you might be left handed and the installer might have reversed the mouse buttons for you. When I reference the left mouse button I am speaking as if the mouse was being used by your right hand.

To click the mouse you press the mouse button down one time and then release the button. This is called clicking because if you listen closely you can hear a soft clicking sound every time the mouse button is pressed.

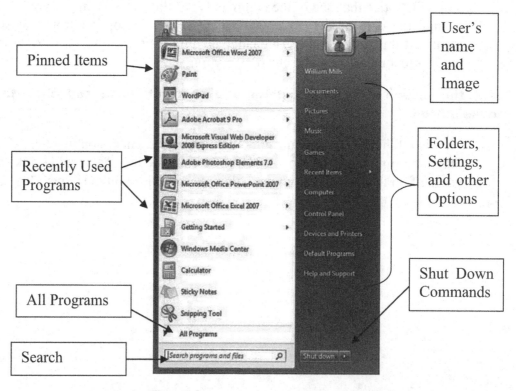

Figure 1-8

14

The Start Menu is made up of a Pinned Items list, a Recently Used Items List, a search box, and links on the right side to folders, settings, and other options. Also on the right side is the user's name and image.

A Pinned item is a link to a program(s) of your choosing for easy access.

The "Recently Used Items" is a list of programs that you can get access to easily. It is similar to the pinned list with one difference, Microsoft put these here not you. This can be changed, well sort of, and will be discussed in Chapter two when we discuss personalizing your computer.

Another change is the Search textbox. This is available at all times when the Start Menu is opened. In the versions before Vista, you had to access the search feature when you wanted to use it. It was not always visible. You use this to search for programs and files on your computer.

You can shut down the computer by clicking the Shut Down button. Clicking the small arrow on the right side of the Shut Down button will allow you to put the computer to sleep, logoff the computer, switch user, or restart the computer.

If the program you want to open is not shown in the Recently Used List, you will need to use the All Programs button to show the rest of the programs. Clicking the All Programs button will bring a list of all the programs that are installed on your computer to the screen. This list will take the place of the opening Start Menu.

Click the All Programs Button

The Start Menu will change and all of the programs available will be displayed in the place of the Start Menu. Figure 1-9 shows the All Programs Menu.

Figure 1-9

Your All Programs should look similar to the one in the figure. You may not have all of the programs that are shown. Some of the programs shown in the figure are for programming and web site development. If the program you want to open is displayed toward the top, all you have to do is click the mouse on it to open it. If the program you want is not shown at the top, such as Solitaire, you may have to click on the folder where the program is located (Games folder for Solitaire) and then click on the icon for the program you want to open.

Would you feel a little better if you tried it?

Move your mouse to the folder that has Games on it

When the mouse pointer touches the folder click the left mouse button once

Slowly move your mouse down the list of programs below the Games folder and as the mouse pointer touches a program the mouse pointer will change from an arrow to a small hand. When the pointer changes into a hand and it is touching the Solitaire program, click the left mouse button once.

The drop down menu showing the Games folder and the programs inside the folder is shown in Figure 1-10.

16

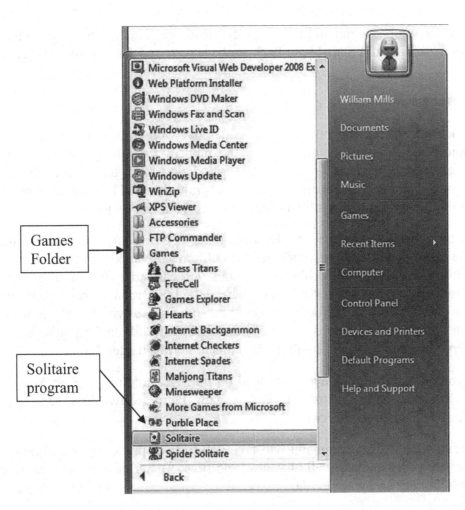

Figure 1-10

The Solitaire program will pop onto the screen, and ready for you to start playing. To remove this program from the screen, you need to "Close" the program. The easiest way to do this is to click the Close button in the top right corner of the screen. Figure 1-11 shows the upper right portion of the screen.

Figure 1-11

The Close button is the small red square with the X in it. Clicking this button will close the open program.

Click the Close button

Click the Start Button again and then move the mouse to All Programs

There is another command that I want you to see. This is the Back command. This will bring the Start Menu back to the screen and remove the All Programs menu.

Click the Back command

You have noticed, by now, that as you move your mouse over the different choices the program, or choice, that the mouse is touching has a highlighted effect on it. This is to let you know that if you click the mouse, this is the command that will be carried out or the program that will be opened. This is just another thing that Microsoft has done to make your life easier.

On the right side of the Start Menu are links to other things in your computer. You can move directly to your documents library by clicking on the Documents Command. There are two library locations you can store files in. There is your private library and a public library. The default location to save your files is your private library. If you were going to share a file with all of the users on your computer, you would probably want to save the file in the public library so everyone could access it. I will explain this more in Chapter 4.

If you click on the Pictures Command on the right side of the Start Menu, you will be taken to the Pictures Library. Again, there are two libraries; a private library and a public library.

Clicking the Music Command will take you to your private music library. This is where you would save all of your music files. Again, there are two libraries; a public library and your private library.

Clicking the Games Command will bring a dialog box showing all of the installed games to the screen. There is a command at the top to see if there are more games available from Microsoft that you can add to your computer. There are over 100 games for you to choose from at the Microsoft game site.

Clicking on the Computer Command will show you the types and storage capacity of the storage devices that are in your computer. Figure 1-12 shows an example of the Computer drives.

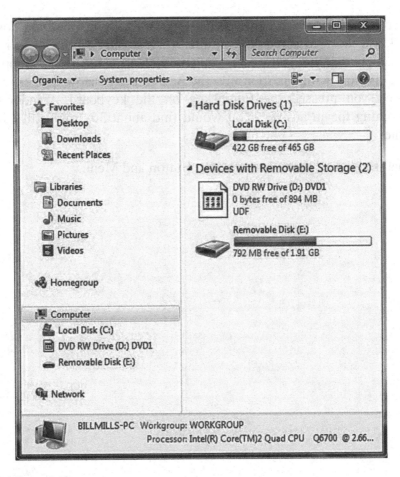

Figure 1-12

The above figure shows that I have a 465GB hard drive on my computer. It also shows that I have a DVD drive and there is also a removable disk attached to my computer. These are not life and death things that you need to know, but I thought that you might like to see what is in your computer.

If you clicked on the Computer Command to see what was in your computer, click on the close button to exit

The Control Panel does what you might think. This brings a new window to the screen that will allow you to make changes to your computer. Most of these you will never have to play with, but there are a few that we will make changes to as we continue through the book.

The Default Programs Command will let you decide which program to use to open different types of files. You should never have to use this section. Windows can usually figure this all out by itself.

The Devices and Printer button (or command as I will undoubtedly use both to reference the same thing) will show you the monitor connected to your computer. It will also show you the printer(s) connect to your computer.

Help and Support is where you can search for things that you don't understand or know how to do. For example: If you didn't remember how to save a document that you just wrote, you could click on the Help / Support button and type the following question into the Search area: "How do I save a document". When you pressed the Enter key on the keyboard, Windows would start searching for an answer. You would find about 30 articles that referenced this topic.

That is about all there is to the Start Button and Menu.

Lesson 1 – 7   The Task Bar

The Task Bar was briefly mentioned earlier and now we will delve into it a little deeper.

Normally the Taskbar is located on the bottom part of the screen. It can, however, be moved to another location on the screen. You can move it to the top of the screen, or the right side, or the left side of the screen. To move the Taskbar one only needs to click on an unused portion of the Taskbar and hold the left mouse button down and drag the mouse to right, left, or top of the screen. The Taskbar will follow the mouse to the new location.

Note: If you try the procedure below and it does not work, the Taskbar is locked and we will discuss that in a few moments.

Move your mouse pointer to an unused part of the Taskbar (probably in the middle) and then click and hold the left mouse button down

Continue to hold the mouse button down and move your mouse pointer across the screen to the right side

When the Taskbar moves to the right side of the screen, release the left mouse button

The procedure you just used is called Drag and Drop. What you did was "grab" the object; in this case the object was the Taskbar, by clicking and holding the left mouse button down. You then moved the mouse to the new location of the object and at the same time you "drug" the object along with the mouse. You then "dropped" the object at the new location by releasing the mouse button. We will use this procedure several times as we progress through the book.

Repeat the above process and move the Taskbar to the top and then to the left side before you put it back at the bottom of the screen.

That was fun, but what if you don't want the Taskbar to be moved. Can you prevent anyone from moving it? I am about to let you in on a well kept secret. You can prevent someone from moving the Taskbar. You prevent the Taskbar from being moved by "Locking" it in place. There are two ways to lock the Taskbar and we will look at the hard way first.

You will probably notice that throughout the book I will first show you the hard way to do something. There really is a reason for doing this, and no it is not because I am mean. If you have to go through all of the steps required to do something, you will probably retain it longer and you will appreciate the easy way a lot more.

Before we can change any of the settings we must bring something to the screen. We need the Taskbar and Start Menu Properties Dialog box. To bring this to the screen we have to right-click on the Start button.

Move your mouse pointer over to the Start button and then press the right mouse button once and then release it

Now click the left mouse button on the Properties choice at the top of the small menu that slides up from the Start button

Now you have the Taskbar and Start Menu Properties Dialog box on the screen. First let's look at the Taskbar part of the dialog box.

Click the tab on the top that has Taskbar on it

Figure 1-13 shows the dialog box after you click on the Taskbar tab.

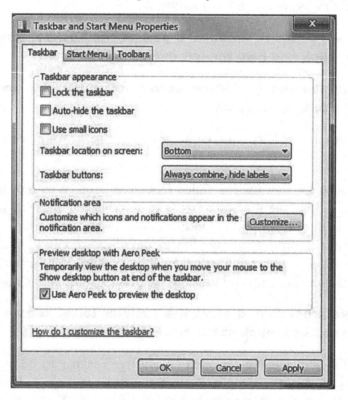

Figure 1-13

The very first choice is "Lock the Taskbar". Next to this is a small square box. This is called a checkbox. That is because if you click this box with the mouse a checkmark will appear inside the box. Now wasn't that clever of Microsoft?

Click the checkbox next to Lock the Taskbar and then click the OK button

Try to move the Taskbar to a new location on the screen

> With any luck the Taskbar should not have moved. If it did move, repeat the above process and make sure that there is a checkmark in the Lock Taskbar checkbox.

> That was the hard way, now for the easy way.

Move your mouse pointer to an unused part of the Taskbar and click the right mouse button

> This will bring a shortcut menu to the screen. This menu is shown in Figure 1-14.

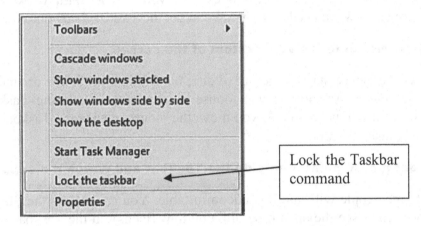

Figure 1-14

> If there is a checkmark next to the Lock Taskbar command the Taskbar is locked. If there is not a checkmark next to it, it is not locked. If you use the mouse to click on the command, the checkmark will toggle on and off.

Click on Lock the Taskbar

Click on it a second time to see the toggle effect

> If, when you tried to move the Taskbar and it did not move, go back and practice moving the Taskbar now.

> Now that you know both the easy way and the hard way to lock the Taskbar, let's see what else we can do from the Taskbar and Start Menu Properties Dialog box.

Right-click on the Start button and choose properties from the shortcut menu

Click on the Taskbar tab to bring it to the front

Under Taskbar appearance, and just below the Lock the Taskbar checkbox, is the Auto-hide the Taskbar checkbox. Clicking this checkbox will tell the Taskbar to hide itself from the monitor until you need it. Obviously Windows 7 cannot read your mind, so when you need it you will have to let Windows 7 know that you need it. This is accomplished by moving your mouse to the area where the Taskbar should be located. When you move your mouse pointer down to the bottom of the screen, the Taskbar will return. When you move your mouse away from the bottom of the screen, the Taskbar will disappear again.

Click the Auto-hide the Taskbar checkbox and then click the OK button

As soon as you click the OK button the Taskbar slipped downward until it was gone from the screen. This does give you a little extra work room on your screen. Now we need to see if we can get the Taskbar back onto the screen.

Move your mouse to the very bottom of the screen

As the mouse starts to go out of site, the Taskbar will slide up and again be on the screen. As long as you mouse pointer stays inside the Taskbar area the Taskbar will be visible. If you move the mouse outside the Taskbar, the Taskbar will again disappear.

Move your mouse pointer outside the Taskbar and watch it disappear

Some people will find this uncomfortable. You cannot see the Start button and you can't see the open program's buttons (icons) on the Taskbar. If you prefer, as I do, to see these things, I recommend that you do not use the Auto-hide feature.

Use the steps above to bring the Taskbar and Start Menu Properties Dialog box back to the screen and make sure the Taskbar is showing in the front

Uncheck the "Auto-hide the Taskbar" checkbox

The next checkbox is the Use small icons checkbox. This will change the size of the icons on the Taskbar so that they are not as large. By default, Windows 7 uses the large icons on the Taskbar.

Click on the Use small icons checkbox and then click OK

Everything on the Taskbar is now smaller. For me this is harder to read and it just doesn't look right. I am going to change mine back to the larger icons. If you feel the same way, you can change yours back as well.

Bring the dialog box back and uncheck the Use small icons checkbox

The next option is to tell Windows 7 where to locate the Taskbar. Before you tell me, I realize that I violated my own rule about showing you the hard way to do something first, but I was in a hurry. I will try not to let it happen again.

By default, the Taskbar is located at the bottom of the screen and you can see this from the dialog box. You can however change this by clicking on the small down arrow on the right side of the list box and then clicking the mouse on the desired location. See figure 1-15 for the Taskbar location part of the dialog box.

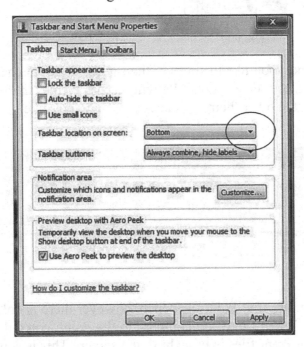

Figure 1-15

Change the location of the Taskbar by choosing the different locations and then bring it back to the bottom

The last thing in the Taskbar appearance section is the Taskbar buttons choices. At first this seems a little confusing, so I will explain the different choices. By default, the Always combine and hide labels choice is used. Let's say that you had two letters that you were writing and you were using Microsoft Word to write the letters. By default, you would see the icon for Microsoft Word and that is all that you would see. Figure 1-16 shows this choice.

Figure 1-16

From this view you cannot tell how many documents are open or which documents they are. The only to tell how many documents are open and the names of the open documents is to bring the mouse to the icon for Microsoft Word and let it hover there for a few seconds. After a few moments the number and names of the documents will be shown directly above the icon. Figure 1-17 illustrates this.

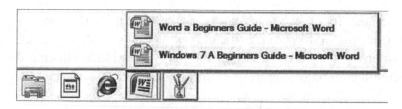

Figure 1-17

Your screen may not look like the ones in the figures. I had to change the theme I was using to get these screen shots. We will discuss themes in the next chapter. If you have one of the other themes, your screen may look like the one in Figure 1-18.

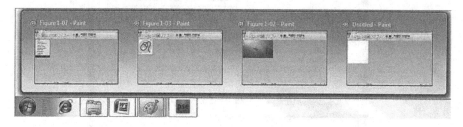

Figure 1-18

In figure 1-18 there are four Paint files open, however there is only one icon for the paint program. When I moved the mouse to the Paint icon a thumbnail view of each of the open Paint files is brought to the screen. This is something that is new in Windows 7. If there are several files open (more than one) you can also click on the icon to see the different files.

If you changed the choice to either of the other two choices (besides the Always Combine and Hide choice) you would not only see the icon but you would also see a label with the partial name of the document. Not only that but you would see a second icon for the second letter and it would also show part of the document name. This way if you wanted to switch back and forth between the documents, you could click the mouse on whichever document you wanted, instead of clicking on the icon and then choosing which document you wanted to bring to the screen.

If you click on the icon for the program that is showing on your screen, the program will minimize (that means that it will shrink down until you can't see it any more). Clicking on an icon for a program that is minimized will cause it to come back onto the screen.

My personal favorite is the second choice; combine when the Taskbar is full. I am going to have you change yours to this, but you can put it back if you don't like it. Figure 1-19 shows the screen view with this choice.

Figure 1-19

Click the down arrow in the Taskbar buttons section and choose the combine when the Taskbar is full choice and then click the OK button

Now whenever you have a program open, you will see the icon for the program and a label next to it. Like I said if you don't like this, you can change it back to the first choice. There is one other thing you should notice. One of the Word document icons is darker and the icon seems to be pressed. This is the active document. This is the one you are working on. The other icon doesn't seem to be pressed and is a lighter color. This is how you tell which one you are currently working on.

Next we have the Notification area. This is located on the far right end of the Taskbar. This section will let us determine which icons are visible in this area. We can customize this area to show the icons that we want displayed. If you want, you can click on the Customize button and **look** at the settings, but **don't** make any changes. It is not that you will hurt anything; it is only that these are the settings that you will want. The Notification area is shown in Figure 1-20.

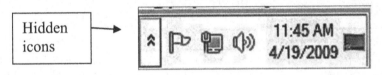

Figure 1-20

There are some hidden icons in this section. If you want to see them, click on the two small arrows (chevron) pointing upward on the left side.

Figure 1-21 shows the hidden icons.

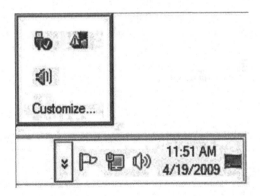

Figure 1-21

27

The last thing on the Taskbar tab is the Desktop Preview. This is also something new with Windows 7. At the far right end of the Taskbar is a small vertical bar. This is the Desktop Preview bar. This is shown in Figure 1-22.

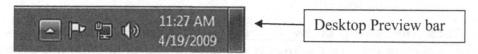

Figure 1-22

If you move your mouse to the Desktop Preview bar, all open programs will become transparent and the desktop will be seem through them. When you move the mouse pointer off the Desktop Preview bar the programs will return to their regular state. That is cool!

What else can we do to the Taskbar? I am glad you asked. We can add toolbars to the Taskbar.

Bring the Taskbar and Start Menu Properties Dialog box back to the screen

Click on the Toolbars Tab

The Toolbars tab is shown I Figure 1-23.

Figure 1-23

28

When I first saw these I thought "this is silly, why would I want these on the Taskbar"? After I tried it out, I thought "Wow, my life will be a lot easier with these". The first one I put on was the Desktop. I know, you have a desktop and this doesn't make a lot of sense. Let's say you are working with a program. Just for fun we will pretend that we are working on a spreadsheet (we pretended to be writing a letter last time and we don't want to make Excel jealous). Our screen is filled with cells and numbers. It is about this time that we remember we were going to restore a file from the Recycle Bin. We could minimize the running program and find the Recycle Bin icon on the desktop and double-click on it to open the Recycle Bin. Or we could use the Desktop tool to immediately access the Recycle Bin. Let's try it and see just how easy it is.

Click the checkbox next to Desktop and then click the OK button

The Taskbar will reflect this change immediately. On the right side, next to the Notification area is the Desktop Toolbar (See Figure 1-24).

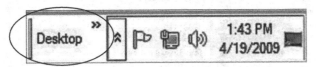

Figure 1-24

You will notice that there are two small arrows pointing to the right. If you click the arrows a list of all of the shortcuts is displayed. To access any of them all you have to do is click on the one you want to access. If there is more than one choice for you to choose from another menu will slide out to the side and you can pick from that menu.

Click on the two arrows in the top right side of the Desktop toolbar section

Figure 1-25 shows the results of clicking on the chevron.

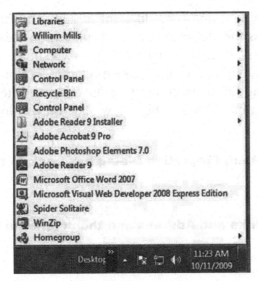

Figure 1-25

29

As I said some of the choices on this menu have other choices embedded inside them. One example of this is the Control Panel choice. If we move our mouse over to the Control Panel choice another menu will slide out to the side.

Move the mouse pointer over to the Control Panel choice

A new menu will slide out from the Control Panel choice and allow you to choose what part of the Control Panel you want to access. Figure 1-26 shows the slide out menu.

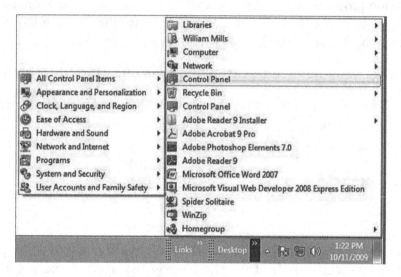

Figure 1-26

Click the mouse anywhere outside of the menu choices and the menu will disappear

I hope you can see that this could save you some time while you are working on your computer. Let's take a look at the next Toolbar you might want to add.

The Links Toolbar is the next one I would want to add. The Link tool bar is fantastic. It lets you access the websites that are displayed on the Favorites Bar. In a like manor, the Address Toolbar will allow you to type in a web address and go directly to the web site, without having to open Internet Explorer. I know that we have not covered using the internet and your favorite websites yet; we will cover them in Chapter six, but let's add these now and you will love them when we get to Chapter six.

Bring the Taskbar and Start Menu Properties Dialog box back to the screen

Click on the Toolbars Tab

Click the checkbox next to Links and Address and then click the OK button

The two new toolbars will be added to the Taskbar and they are shown in Figure 1-27.

Figure 1-27

How to use the last two toolbars will be covered in Chapter 6 when we discuss using the internet.

The last tab on the Taskbar and Start Menu Properties Dialog box is the Start Menu Tab. There is not much on here but we need to look at it.

Bring the Taskbar and Start Menu Properties Dialog box back to the screen

Click on the Start Menu Tab

The first thing you should know is that there is a proper way to shut Windows 7 down. All of the programs need to be closed and the computer needs to shut down in an orderly fashion. You do not just push the On/Off button to turn the computer off. If you turned the power off to shut the computer down you would end up with corrupted files and damaging your computer. No matter how many times you tell some people, they still want to turn the computer off by pressing the power button. Microsoft is trying to protect you from yourself. You may think that this is pretty arrogant of them, but I imagine that they have had to fix their share of messed up hard drives and this will possibly save you a headache.

Figure 1-28 shows the Start Menu Tab.

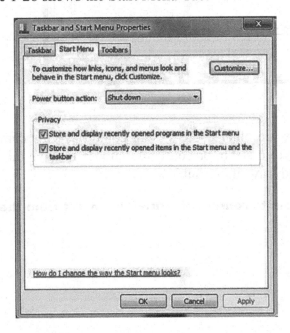

Figure 1-28

We can decide what should happen if we press the power button. By default Microsoft has decided to make the computer shut down in an orderly fashion if you press the On/Off button. This means that the computer will close the open programs and shut down the way it is suppose to shut down, it won't just turn the power off. This is normally what you will want to have this set at, but you can choose a different setting.

Click the down arrow across from Power button action and see the other choices

Figure 1-29 shows the choices.

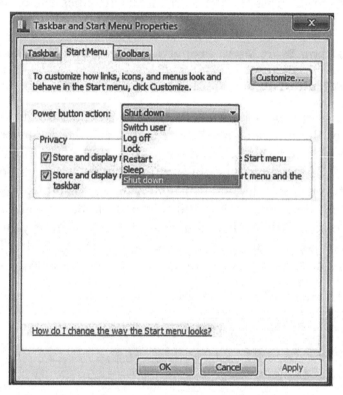

Figure 1-29

As you can see there are other choices. You could have the computer just log you off, or perhaps go to sleep, or you could choose to have this action restart the computer. I would leave it on shutdown.

Click the down arrow again to remove the drop down list from the screen

We talked earlier about the Start Menu having the recently opened programs on it. That is true, but you get to decide if the recently opened programs are allowed to be on the Start Menu. Also you can choose to have the recently opened programs appear in the taskbar. These choices are in the Privacy section. I would leave these choices as they are. The only one I might change is the one that has the option for the recently opened items to be displayed in the Taskbar. I think you will like having this option on. To toggle the choices off and on simply click the mouse on the corresponding checkbox.

This part is different than the choices in the Privacy section. It deals with the recent files that you have had opened. You can add a button to the Start menu that will give you quick access to the recently opened files. These might be letters that you have written or perhaps spreadsheets that you were using.

Here is how you get the recently opened files in the Start Menu.

Click the Customize button by the top

The Customize Start Menu Dialog Box will come to the screen as shown in Figure 1-30.

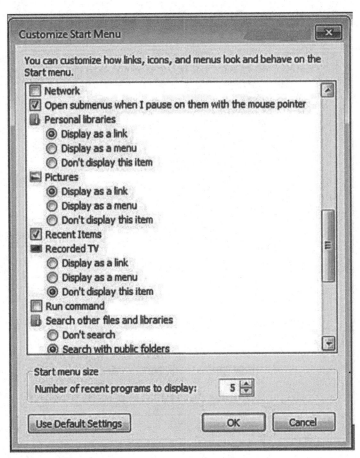

Figure 1-30

Scroll down until you see Recent Items and then click the checkbox next to it

Click the OK button to close this screen

Click the OK button to close the Taskbar and Start Menu Properties Dialog box

Now when you open the Start Menu you will find a new command (probably just below Games on the right side). This is the "Recent Items" command. A list of the recently opened files will show up when you click this button. To open one of the files, click on it with the mouse.

Lesson 1 – 8   Finding a Program

In this lesson we will talk about how to find a program on your computer. As usual there is more than one way to find a program. This lesson will cover the most common ways to find a program.

The first way is by using the Start Menu. This is the most common method of finding and opening a program. Since we are talking about the basic things you need to know we will find, only find, the program in this lesson. In Chapter three we will open and close programs.

In this lesson we will be looking for the Calculator program. This simple calculator comes with Windows 7 and every computer will have this program.

The first and most basic rule to remember is all things start with the Start button.

Click the Start Button

This will bring the Start Menu to the screen (See Figure 1-31).

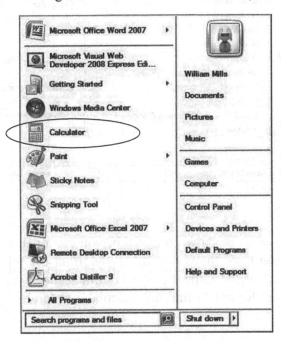

Figure 1-31

If you are real lucky (and you have had the program open recently) it might be on the Recently Used Programs List and visible on the Start Menu. If it is, it can be opened and you can work with it.

Well that was easy. Hey, what if the program is not visible on the Start Menu?

There is a good chance that the program is listed in the All Programs section of the Start Menu. Let's see if it is there.

Move your mouse over the All Programs command and let it hover there for a few seconds

In a few moments the Start Menu will change to reveal the other programs and folders available. This is shown in Figure 1-32.

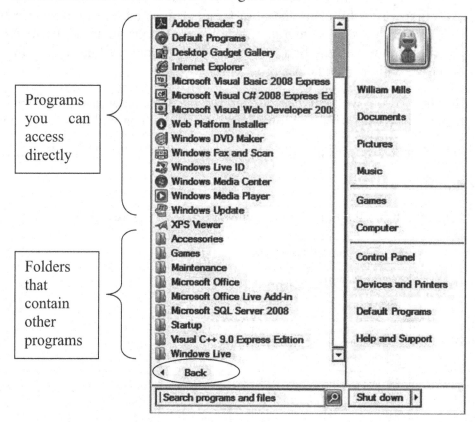

Figure 1-32

If the program that you want is shown at the top, you can open it with just the click of the mouse. I don't see the Calculator program at the top. That means that it must be inside one of the folders in the bottom section. We may have to search through a bunch of folders to find the Calculator program. There must be an easier way that that.

Move your mouse to the Back part of the Start Menu

In a few moments the original screen of the Start Menu will be back on the screen. At the bottom of the menu is the Search programs and files textbox. You can type the name of the program you are looking for and let Windows 7 find it for you.

Using the keyboard type Calculator in the textbox

The insertion point (the flashing vertical line) is already inside the search textbox. You do not have to click the textbox with the mouse or anything, just start typing.

As you type, Windows 7 will start looking for programs, files, and folders that match what you are typing. When you are finished typing Windows 7 will show you the results from the search. My results will probably be different from your results. Mine will show a reference to the calculator picture I have on my computer. Your search results will probably only show the one program that is on your computer and references to the Gadgets which will be discussed in another chapter. Figure 1-33 shows my search results.

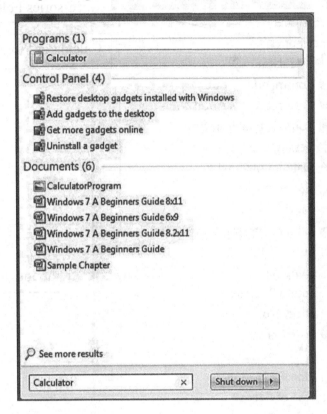

Figure 1-33

At the very top you can see that the search found one program with that name. If we want to open the program all we have to do is click the name with the mouse.

If you want to see where the program is located, you will need to right-click on the program and choose properties form the available choices. The location of the program is shown under the General Tab. The program is located at the following location:

C:\Program Data\Microsoft\Windows\Start Menu\Programs\Accessories

Click the mouse anywhere outside of the search area to close the results window

I am going to tell you where the program is located so we can go and actually find the program in the Start Menu. The Calculator program is located in the Accessories folder.

Bring the All Programs part of the Start Menu back to the screen

Click the Accessories folder with the mouse

The folder will open and show you its contents.

Figure 1-34 shows the folder contents.

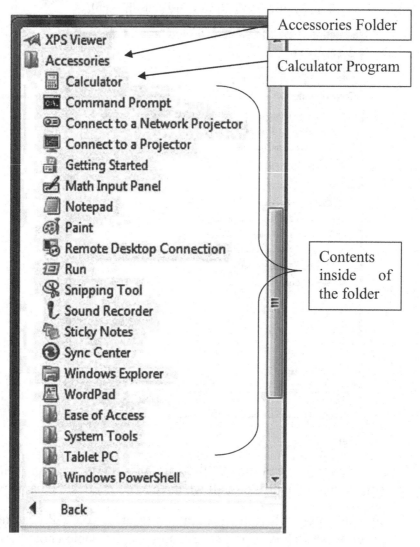

Figure 1-34

To access the program all we have to do is click on it.

Click anywhere outside of the menu area to close the Menu without selecting anything

If you don't know where the program is located and you don't want to start looking through the folders to try to find it, the easiest way to find it is to run the search.

Lesson 1 – 9   Pinning a Program

Pinning a program sounds like a strange name for a lesson, but I think that you will find this lesson very useful. Earlier we talked about the Start Menu having a Pinned items section. In this lesson you will learn how to Pin an item to the Start Menu.

This should not be a real long lesson. This is because pinning an item to the Start Menu is a very simple task. In this lesson we will pin the WordPad program to the Start menu. WordPad is a simple word processor program that comes as part of Windows 7. You can use this to write letters and other such things. It is not as powerful as Microsoft Word, but it is better than Notepad, which is the first word processor that Microsoft provided way back in Windows 3.1. At least that is the first that I remember.

Open the Start Menu and select All Programs

Click on the Accessories folder to open it

Look down the list of programs and locate the program called WordPad, but do not click on it

Figure 1-35 shows the way to get to the program.

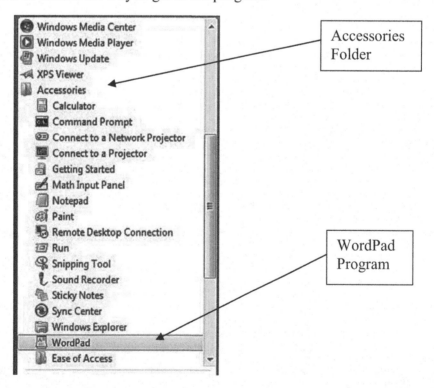

Figure 1-35

If we just click on the program it will open, but this is not what we are going to do in this lesson. In this lesson we want to add this program to the top part of the Start Menu. Figure 1-36 shows the pinned part of the Start Menu.

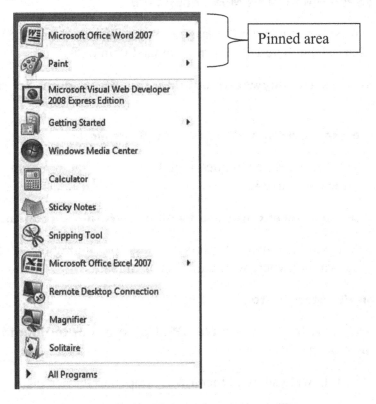

Figure 1-36

When we add this program as a permanent part of the Start Menu, at least until we change our mind, it is called pinning the program to the Start Menu.

Move your mouse to the WordPad program and right-click the mouse

The following shortcut menu will appear.

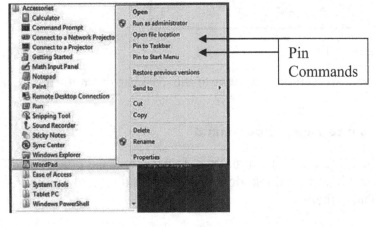

Figure 1-37

41

If we want the program to be on the Start Menu or the Taskbar at all times we need to click on the appropriate pin command.

Click on the Pin to Start Menu command

The shortcut menu will disappear from the screen and the next time we open the Start Menu there will be a new pinned item on it.

Click the mouse anywhere outside the Start Menu to remove it from the screen

Click the Start button and look at the Start Menu

There will be a new pinned item by the top. The program will be there until you decide to remove it.

We will use a different process to pin the WordPad program to the Taskbar

We could have used the same process for pinning the WordPad program to the Taskbar, but where would the fun be in that?

Click on the Start button

Move your mouse up to the WordPad program in the pinned item section and right-click on it

The following shortcut menu will jump onto the screen.

Figure 1-38

From here we can pin the item to the Taskbar or unpin the item from the Start Menu.

Click on the pin to Taskbar command

Now both the Start Menu and the Taskbar will have the WordPad program on them at all times. Look down at the Taskbar and make sure the icon for WordPad is there.

Unpin the program from both the Taskbar and the Start Menu by right-clicking on each item and choosing the unpin command

Armed with this knowledge, you can add any program to the Start Menu and/or the Taskbar.

Lesson 1 – 10   Adding Shortcuts

Did you ever wonder how those icons for the different programs got on the computer screen on your desktop? This lesson will teach you how to add these shortcuts to your desktop.

In this lesson we will add the WordPad icon to the desktop and create a shortcut to get to the program quickly.

Click on the Start button and find the WordPad program

Remember, the WordPad program is in the Accessories folder. Once you find the program you will need to right-click on it.

Right-click on the WordPad program

The same shortcut menu will appear just like when we pinned the program to the Taskbar and Start Menu. This time we want to move the mouse over to the "Send To" option. When you do another menu will slide out to the slide. From this menu we will choose the Desktop choice.

Move the mouse to the "Send To" choice and then click on the Desktop (Create shortcut) choice

Figure 1-39 shows what the menus look like.

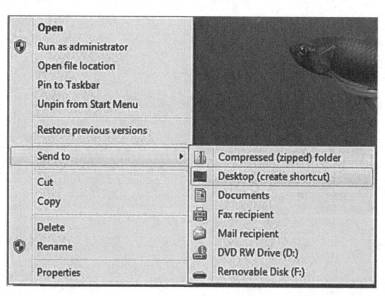

Figure 1-39

When you click the Desktop choice, you may not see anything happen on the screen. This is because the Start Menu is covering most of the left side of the screen. You may have to close the Start Menu to be able to see the new shortcut.

Click anywhere outside of the Start Menu to remove it from the screen

You should now be able to find the new shortcut on the desktop. Figure 1-40 shows you what the new shortcut should look like.

Figure 1-40

You can now add shortcuts for all of your favorite programs. In the past few lessons you have learned how to add a favorite program to the Taskbar, the Start Menu, and now the Desktop. This should make your life a little easier.

Lesson 1 – 11   Mouse Settings

I did not include this lesson with the other lessons on the mouse, because this lesson is different than the other lessons. The other lessons were the basics things that you can do with the mouse. This lesson will deal with the mouse settings and how the mouse behaves.

For this lesson, we will be using the Control Panel. The control Panel is where you adjust your computer settings. We want to be somewhat careful when using the Control Panel as there are a few places that we do not want to make any changes. We will come into the Control Panel a few more times before we are finished, but not too many.

Click on the Start Button

Click where it says Control Panel on the right side about half way up

The words Control Panel will turn a different color when the mouse is positioned on it and then you can click the mouse. The Control Panel will jump onto the screen. This screen will look different than what you are used to seeing if you are familiar with some of the earlier versions of Windows. Figure 1-41 shows the Control Panel screen.

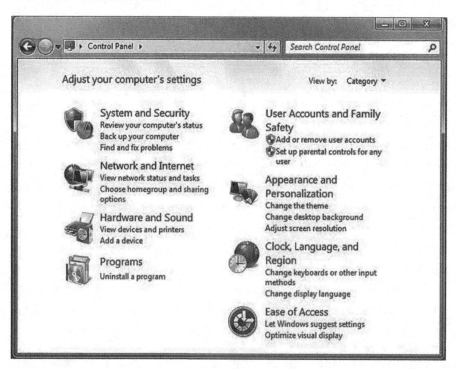

Figure 1-41

The easiest way to find the mouse settings is to click on the small black arrow pointing to the right directly after the words Control Panel.

Click on the small downward pointing arrow to the right of the words Control Panel

Figure 1-42 shows the drop down menu list that appears when you click the black arrow.

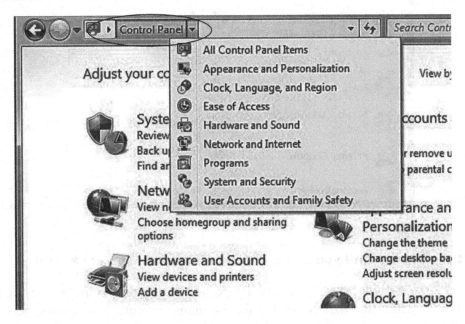

Figure 1-42

Click on All Control Panel Items

This will bring another window to the screen. This is the All Control Panel items screen, bet you didn't see that name coming. This window is shown in Figure 1-43.

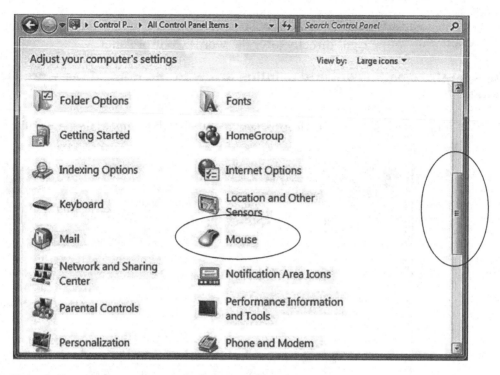

Figure 1-43

Scroll down until you see the Mouse icon and then click on it

This will bring the Mouse Properties Dialog box to the screen. This is where we will customize the setting for the mouse. Figure 1-44 shows the Mouse Settings Dialog box.

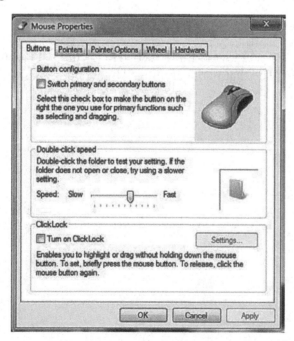

Figure 1-44

Remember when I said if you were left handed and the installer might have changed the buttons around for you. This is where you would change the buttons. The checkbox under Button Configuration, if checked, will reverse the two buttons on the mouse.

The next section is the Double-click speed section. This is where you can change the timing if you are having trouble getting the "double-click" to work. In the very center of the dialog box is a slider that is used to adjust the speed of the "double-click" function. Directly to the right of the slider is a small folder inside a box. This is the practice area. You can practice double-clicking the mouse on this folder and see if you can open it. If you have trouble opening it, try adjusting the slider (probably to the left to slow the speed down).

Double-click on the practice folder and try to open it

Adjust the speed a little slower or faster until you find a speed that will let you open the folder when you double-click the mouse

If you find a speed that you like, click the Apply **button at the bottom**

The last section on this tab is the ClickLock section. Later when we are selecting text and moving it around inside of a document, you may want to come back to this section of the Control Panel and change this setting. When you are moving an object around on the computer screen, you must hold the mouse button down while you are moving the object. If you turn the ClickLock feature on you won't have to hold the mouse button down. I will try to remember to remind you of this later.

If you get tired of the mouse pointer being a small white arrow you can change it to something else on the Pointers Tab. Let's look at some of the options available.

Click the mouse on the Pointers Tab (See Figure 1-45)

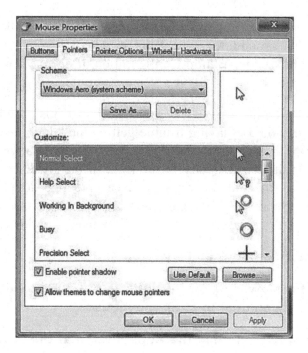

Figure 1-45

The scheme section at the top is where you can change the mouse pointer to a different type. You can click the down arrow and then click on the available choices and watch the bottom area to see what the different choices look like.

Note: Before you click the down arrow, notice the scheme you are currently using.

Click the down arrow in the Schemes section and then click on some of the available choices to see what they look like

When you are finished playing change the scheme back to the original scheme

If you have forgotten the original scheme it was probably Windows Aero (system scheme).

You may have trouble getting your mouse to settle on an object. It may seem that the mouse moves past the object before you can stop it (an object can be anything on the screen). If this is the case you may want to change the speed of the mouse pointer. You can make the mouse pointer move faster or slower. This action is performed from the Pointer Options Tab.

Click on the Pointer Options Tab

Figure 1-46 shows this tab.

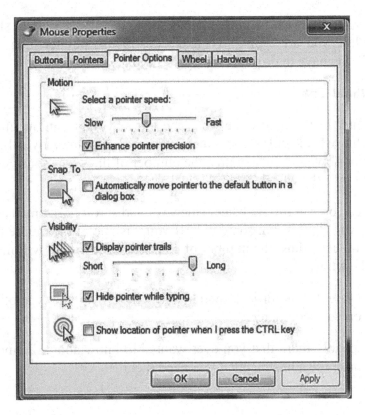

Figure 1-46

Click on the slider under motion and adjust the speed from the center of the slider to the right to make the mouse pointer move faster

Move the mouse around and see the difference the speed change made

Repeat the process only move the slider to the slow side

Find a speed that you feel comfortable with and then click the Apply **button**

If you check the box in the Snap to section, the mouse will move to the default button (usually the OK button) anytime there is a dialog box open on the screen. I would leave this unchecked if it was me.

There are times when I have trouble finding my mouse pointer. I know that sounds silly, but there are times when I move the mouse and I just don't see it on the screen. I keep moving it until I happen to see it, but there is an easier way to help me see it when I move the mouse.

In the Visibility section of the tab is a checkbox to display the pointer trail. Having this checked will let a small trail of mouse pointers follow the mouse as you move it. I prefer to have this checked and have the trail set to long. If you want to see the difference the trail makes; click the checkbox to either turn it off or on and then move the mouse and watch the screen. You can also try adjusting the length of the trail.

51

Another thing that is very annoying is to have the mouse pointer showing on the screen while you are typing. To make the mouse pointer hide while you are typing all you have to do is click this checkbox.

Click on the Wheel Tab

In the center of the mouse, between the left and right buttons, may be a scroll wheel. This wheel will let you move the mouse pointer by rolling the wheel either up or down. You can adjust the number of lines the mouse will scroll each time you move the wheel.

Sometimes you can tilt the wheel sideways and scroll horizontally. Not all types of mouse will do this, but some will. If yours is one of the mouse types that will do this, you can adjust the number of characters the mouse will move each time you tilt the wheel.

The Hardware Tab is for information and will tell you the type of mouse that is installed on your computer.

This is probably more that you ever wanted to know about the mouse, but now you can make the adjustments if you need them.

Lesson 1 – 12   Using the Keyboard

Now that you have mastered the mouse, it is time to move on to the other device that you use to control your computer, the Keyboard. The keyboard may seem more familiar to you and therefore you could assume that it is easier to use than the mouse. Well, we will see. There are some extra keys on the keyboard that you may not be familiar with, so in this lesson we will cover those extra keys and what they are used for. Figure 1-47 shows a picture of a common keyboard.

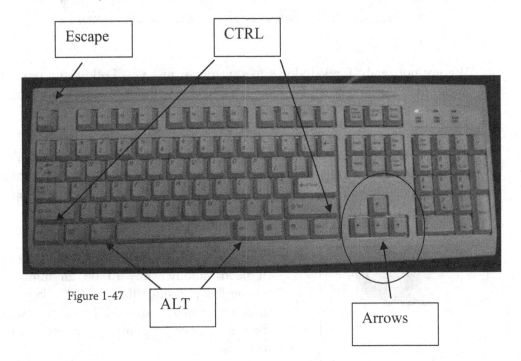

Escape   CTRL

Figure 1-47   ALT   Arrows

Press and hold the <ALT> key, press the <F4> key, and then release both keys.

Pressing these keys commands the current program to close. Since you are using the Windows Desktop, the Shutdown Windows dialog box appears, as shown in figure 1-48

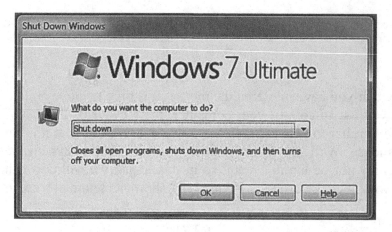

Figure 1-48

We are not ready for Windows to shut down just yet. Follow the next step to back out of the shut down Windows dialog box without selecting anything.

Press the <Esc> key.

The ESC (Escape) key is located on the top left side of the keyboard. Pressing the Escape key does the same thing as clicking the Cancel button. The Shutdown Windows dialog box disappears and we are back at the Windows Desktop.

Table 1-3 lists the special keys and their functions.

	The <ALT> key doesn't do anything by itself; it needs another key to do anything. For example, pressing the <TAB> key while holding down the <ALT> key, switches between any programs that are currently running.
	Just like the <ALT> key, the <CTRL> key doesn't do anything by itself; you need to press another key along with it to do anything. For example, pressing the <X> key while holding down the <CTRL> key, cuts whatever is selected.
	The <F1> key is the Help key for most programs. Pressing it displays helpful information about what you are doing and can answer about the program.

	The <ESC> (Escape) key does the same as clicking the Cancel in a dialog box. For example, if you click something and an unfamiliar dialog box opens, you can close it by pressing the <ESC> key.
	The <Enter> key is has the same function as clicking OK in a dialog box. For example, you can type the name of a program you want to run in a dialog box, and then press <Enter> to run the program. If you are entering text, The <Enter> key adds a new line or starts new paragraph.
	When you are in a dialog box, or spreadsheet, pressing the <TAB> key moves to the next field. If you are using a word processor, the <TAB> key works just like you'd think it would: it jumps to the next tab stop whenever you press it.
	The arrow keys will move your computer's cursor on the screen if you are in a text document. They can also move you to the next selection in a dialog box.
	Nothing surprising here. The <Delete> key erases or deletes whatever you select – files, text, or graphical objects. If you are working with text, the <Delete> key erases characters to the right of the insertion point.
	Use the <Backspace> key to fix your typing mistakes – it erases characters to the left of the insertion point.
	The <Home> key jumps to the beginning of the current line when you are working with text.
	The <End> key jumps to the end of the current line when working with text.

	The \<Page Up\> key moves up one screen.
	The \<Page Down\> key moves down one screen.

Table 1-3

There are other keyboard shortcuts we will deal with a little later.

Lesson 1 – 13   Setting the Time & Date

We mentioned earlier that you could set the time and date on your computer. In this lesson I will show you how to change the settings for the time and date.

Earlier I mentioned that this could be done by right-clicking on the time in the Notification area. This is still true, but I want to show you another way.

Open the Control Panel like we did when we changed the setting for the mouse

Locate the Date and Time choice and click on it

It should not be difficult to locate this as the different things that you can do are in alphabetical order. When you click on it, the Date and Time Dialog box will come to the screen as shown in Figure 1-49.

Figure 1-49

If you want to change the date and/or time, simply click on the Change date and time button.

Click on the change date and time button

The following dialog box will come to the screen.

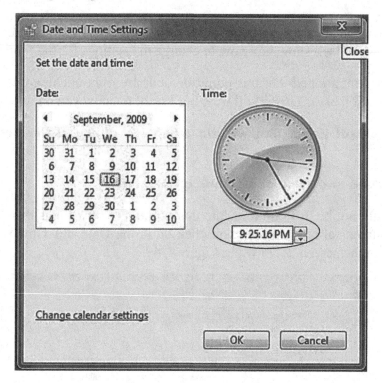

Figure 1-50

The arrows to the left and right of the text showing the current month will move the calendar to the previous or next month. If you need to change the date to a different day all you have to do is click the mouse on the date you want.

If you want to change the current time you need to use the up and down arrows to adjust the time. If you want to adjust the hour, you can click on the up and/or down arrows. If you do not select a specific part of the time to adjust (hour, minute, or second), the arrows will change the hour. To adjust the minute you can click the mouse on the digits that represent the minute and then use the up and down arrows. The second's adjustments work the same way.

Note: If you don't feel comfortable clicking the mouse on the minutes or seconds section, you can press the TAB key to move from the hour to the minutes section, and from the minutes section to the seconds section.

If the time shown in the dialog box is correct, press the Cancel button. If the time is not correct, set it to the correct time and then click the OK button

This will bring you back to the screen with the Date and Time Dialog box on it.

If, for some reason, when you received your computer the time zone, shown in Figure 1-49 in the center of the dialog box, does not have the correct time zone shown, you can change the time zone to match the one you are in.

Click the Change time zone button

This will bring the Time Zone Settings Dialog box to the screen as shown in Figure 1-51.

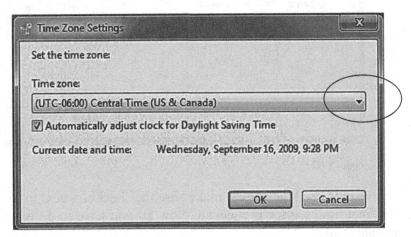

Figure 1-51

If the current setting is not correct, you can click the down arrow and scroll down the list until you find the correct time zone and then click on it. After that you will need to click the OK button to set the time zone.

Also you should notice that the checkbox next to "Automatically adjust clock for Daylight Savings Time" is checked. You will want to leave this checked.

If you made any changes, click the OK button. If not, click the Cancel button

This will bring the Date and Time Dialog box back to the screen. At the top are three tabs: Date and Time, Additional Clocks, and Internet Time.

Click on the Internet Time Tab

The last thing you might want to do is make sure your computer clock synchronizes with the internet. The internal clock on your computer can run fast or slow (I have seen both). If the screen on the dialog box does not say "This computer is set to automatically synchronize with" and then the name of the service you will synchronize with, click on the Change Settings button. Oh well, let's just do it anyway.

Click on the Change Settings button

A new dialog box will come to the screen. This time it is called the Internet Time Settings Dialog box. It is shown in Figure 1-52.

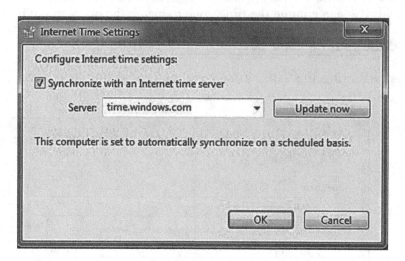

Figure 1-52

If you want to use this feature, make sure the checkbox next to Synchronize with an internet time service is checked. Now all you have to do is select the service you want to use.

Click the downward pointing arrow after Server to see the available choices

There are about 4 government services that are available and there is always the Windows time service. Windows will probably work, so you can choose that service.

Click on time.windows.com

You will notice that my dialog box says that this service is not running. You can start the service for this by clicking the Update button. When you click the Update button, it will tell you that an error occurred and it cannot update. When you click the OK button the service should start in a few moments.

Click the Update Now button

Click the OK button

Now you know most everything you will need to know about setting the time on your computer.

Lesson 1 – 14 Exiting Windows

When you are finished using your computer, you need to shut down Windows before you "turn off" your computer. Shutting down gives Windows a chance to "clean things up" and close the open programs running in the background, the things you never see. It also gives Windows a chance to save any information in the computer's memory to the local disk, cleaning up temporary files, and verifying that you've saved any changes that you have made to any files you have worked on.

Always follow these steps when you shut down your computer:

Save all your work and exit all your programs

Saving any files you have been working on is the most important step of all when you shut down your computer.

After you have closed and saved any open files; click the Start button.

The Start menu will appear. Directly across from the search textbox is the Shutdown button (see Figure 1-53).

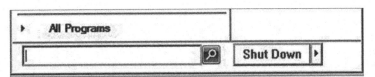

Figure 1-53

In previous versions of Windows if you clicked the Turn off computer button you were given a few different choices. These included switching the user, logging off, and restarting the computer. If you click the Shutdown button the computer will shutdown.

The other choices are still available to you but you have to click the arrow to the right of the shutdown button. Figure 1-54 shows the results of clicking the arrow.

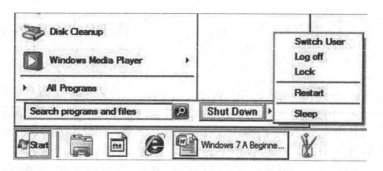

Figure 1-54

Click the Shutdown button and the computer will turn off

You will need to turn your computer back on to continue

Chapter Two   Personalizing your Computer

You are the Master of your domain, the king (or queen) of your castle, and this is YOUR computer. You want your computer to be different that the average computer that someone brings home from the store. You want your computer to reflect your personality. You want to personalize your computer.

This chapter will teach you how to make this "your" computer.

Lesson 2 – 1   Desktop Background

To make your desktop more pleasant to look at, Windows covers it with pretty pictures known as a background. Some people refer to the background as wallpaper. Tired of Windows 7's normal background? You can change the picture used for it. You get to choose your own picture, any picture stored on your computer. It could be a picture of your cute, cuddly kitten or even your children or grandchildren. To change the background, follow the steps below.

As you might expect there is more than one way to get to the same place in your computer. We will look at the hard way first. You better go get a cold soda, this is a tough one.

Click the Start button

Click "Control Panel" on the right side

Click on Appearance and Personalization

Click on Personalization

Wow, I told you that this was going to be tough.

The easy was to right-click on an empty part of the desktop and then click on Personalize.

On your screen is the Personalization window. This is where you will start to make your computer "your" computer. Figure 2-1 shows the Personalization window.

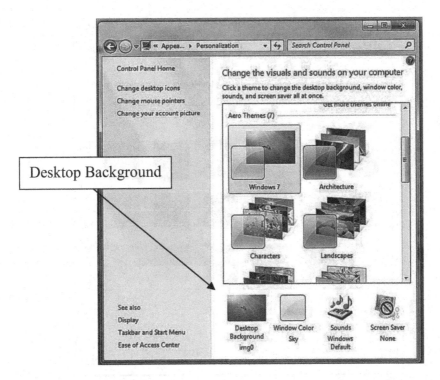

Desktop Background

Figure 2-1

The first thing we are going to personalize is the Desktop Background. I cannot be sure of what your computer screen Looks like at this point. You may have a picture of a fish, or mountains, or perhaps a cow on your screen. We can change the picture on the Desktop by clicking on the Desktop Background icon (where the arrow is pointing). Whatever you currently have as your background will show on this small square.

Click on Desktop Background

If you have the background that does not have a picture and has only a solid color as the background, the choices that come to the screen will only be a choice of colors for the background. This is shown in Figure 2-2.

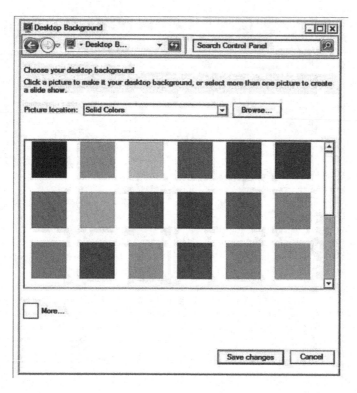

Figure 2-2

If you have a picture on your background, the results of clicking the Background button will be quite different. You will be presented with a choice of pictures that you can use for the background. The screen for some of the pictures available is shown in Figure 2-3.

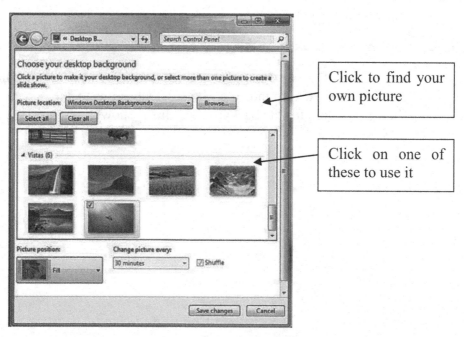

Click to find your own picture

Click on one of these to use it

Figure 2-3

If you like one of the pictures from Windows, you can click on it and the background picture on your monitor will change to match your selection.

If, on the other hand, you want to use a picture that you have stored on your computer's hard drive, you can just as easily change your background to your picture.

To use one of your own pictures, click on the Browse button

The Browse for Folder Dialog box will come to the screen (See Figure 2-4).

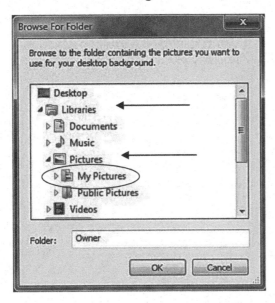

Figure 2-4

More than likely your pictures will be in the My Pictures folder, which is in the Pictures group under Libraries. To get to the My Pictures folder, click the small white arrow to on the left side of Libraries. Then you click on the small white arrow on the left side of Pictures. Next you need to click on the My Pictures folder. Finally you need to click the OK button. Now you will see your personal pictures that have been stored on the hard drive. When you find the picture that you want, click on it and then click the Save Changes button. The picture on your desktop will change to match the picture you have chosen.

That sounds like a lot of work to change the background picture on your computer. But now you will be able to look at a picture of something special every time you use your computer. This could be a picture of your children, grandchildren, your pet, your girl friend (or boy friend), or you could be like me and have your own picture on the screen.

If you have made any changes and you are happy with them, click the Save Changes button, otherwise click the Cancel button

We need to back up for another "what if" question. What if you would like to have a picture on your desktop but you don't always want the same picture on it? In lesson 5 we will be discussing Themes and we will also show you how you can have a slide show as a background. That is going to be fun.

Lesson 2 – 2   Changing Colors

In this lesson, you will learn how to change the color of your windows, Start Menu, and taskbar. These changes won't make your computer run better and you may decide that you like the choices Windows made for you, but you also might like the other choices that are available. We need to go back to the Personalization screen to make these choices.

Right-click on an unused part of the Desktop and choose Personalize from the shortcut menu

You will be back to your Personalization window (See Figure 2-5).

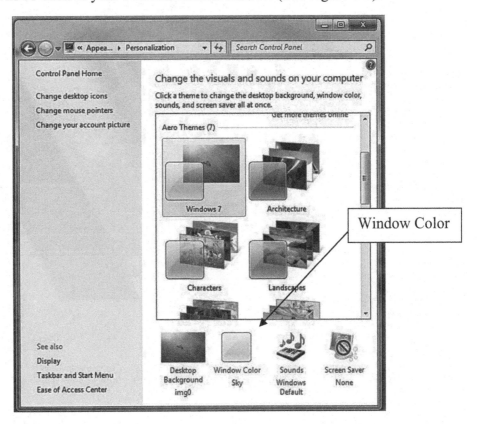

Figure 2-5

Click on Window Color

Your screen will change to show the colors that are available to you. If you see one that you like, you can click on it and click on "Save changes".

Click on each of the choices and watch the Taskbar and the top of the color choice window to see the affect each will have

When you are finished you can click on the "Save changes" button, or the "Cancel" button to not save the changes that you made

Lesson 2 – 3　　Screen Savers

I am sure that you have walked away from your computer and returned to find that your screen has changed into something completely different. There may have been what appeared to be stars whizzing by or perhaps the Windows logo was floating on the screen. This is the screen saver. It was designed to protect your computer screen from having an image permanently burned onto the screen. Way back with cathode ray tubes, like the older televisions had, if an image was left on the screen in one place the image would be permanently etched into the screen. The screen saver would prevent that from happening by changing the image on the screen, so that nothing remained in the exact same place all of the time. We still use screen savers today.

This lesson will teach you how to add, or change, a screen saver.

Bring the Personalize window back to the screen if it is not there

On the bottom right side of the window is the screen saver option. Using this you can add a screen save if you are not using one, or you can change the image displayed by the screen saver.

Click on Screen saver

A new window (Dialog box) will come to the screen. This is the Screen Saver Dialog box and is shown in Figure 2-6.

Figure 2-6

As you can see, I have no screen saver activated on my computer. This is not good; I need to get one going.

Click the down arrow to bring the list of available screen savers to the screen

Figure 2-7 shows the drop down list.

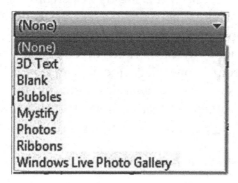

Figure 2-7

Click on each choice and watch the screen directly above the drop down list to see how they look

You might prefer the Photos choice as your screen saver. This will randomly show pictures that you have stored on your computer. This is one screen saver that has some additional settings.

Click on the Photos choice and then click on the settings button

The settings Dialog box will come onto the screen as shown in Figure 2-8.

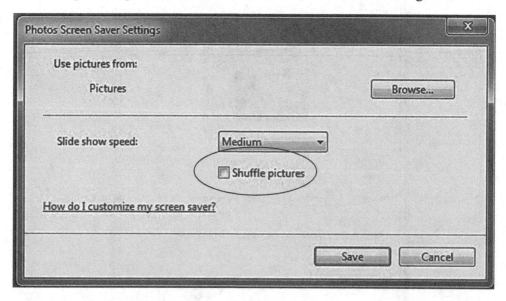

Figure 2-8

By default, Windows will use the pictures that are stored in the Pictures folder. If you want to use the photos that you have saved, and they are not in the Pictures folder, you will need to click on the Browse button and let Windows know where the pictures are located. The same dialog box shown in Figure 2-4 will come back to the screen and you can select the folder that has your pictures in it.

I like the shuffle pictures choice because I never know which picture will enter and leave the screen.

I also like to use the medium speed. It is not to fast or to slow. I have time to see every picture.

When you have decide on the choices you want, click the Save button

Now you will go back to the Screen Saver Settings Dialog box. There are two other things we want to look at. The first choice is how long do we wait for the computer to sit idle before the screen saver starts working. You can adjust this to any number of minutes starting at one minute up to the maximum of 9999 minutes. Anywhere from five to fifteen minutes is normal.

The second choice is what happens after the screen saver starts and you move the mouse or touch a key on the keyboard which removes the screen saver. You can have the computer come back to the screen you were on when it started or you can have the computer go back to the login screen. If you are the only person using this computer, having it go back to the regular screen is fine. If someone else may come over and use your computer, you might want to go back to the login screen. If you set up a password with your screen name, you will have to enter the password before you can go back to working on the computer. If you want to go back to the login screen, make sure the checkbox next to "On resume, display the logon screen" is checked. If you want to just go back to working on your computer without logging back in, leave this unchecked.

When you are finished you can click the OK button to save any changes that you have made

Lesson 2 – 4   Power Settings

The power settings are not something that most people even think about let alone try to adjust. Before you panic, this does not affect anything in your house or anything like that. This only deals with how you can save power and prolong the life of your computer.

Bring the Control Panel to the screen and click the down arrow to the right of "View by"

Click on the choice "Large Icons"

Scroll down and click on Power Options

If you don't remember how to bring the Control Panel to the screen, go back and review lesson 2-1.

The Power Options Dialog box will jump onto the screen (See Figure 2-9).

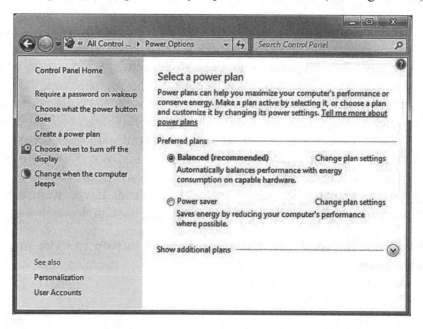

Figure 2-9

On the right side of the dialog box are the power plans. The recommended power plan is the balanced plan. This will allow Windows to balance the performance and power consumption features. It will not reduce the performance of the computer to try to save power, which the power saver plan does. Personally, I would leave this set to "balanced".

From this dialog box you can also decide when your computer should go to sleep and when the display should be turned off. This does not mean that your computer has a curfew and it must be in bed by ten o'clock. This means that if your computer sits idle for an extended period of time (that you get to determine) how long do I wait until the computer starts conserving power and only performs the essential functions to keep it going. In effect, you (well, Windows 7) is putting your computer to sleep. Also to conserve power Windows 7 can turn the monitor screens off at a specified idle time.

Click on "Change when the computer sleeps"

The dialog box will change to the Edit Plan Setting Dialog box. This is shown in Figure 2-10.

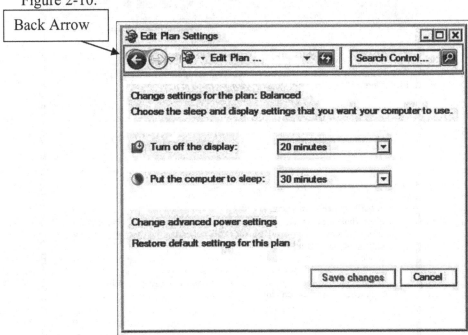

Figure 2-10

The default setting are probably shorter that the ones in the figure. If I remember correctly, the default time to turn off the display is ten minutes. For my computer that seemed a little too quick. I may have to leave my computer for a few minutes, you know, coffee break, and I don't want it to go to sleep that quickly. I set mine for twenty minutes; it doesn't take me twenty minutes to drink my coffee. If I still have not used my computer in thirty minutes, I let Windows put the computer to sleep. If it is sitting there that long just waiting on me to do something, it needs to go to sleep (I probably have).

If you have made changes to your computer's settings, click the "Save changes" button

If you have not made any changes, click the "Back arrow"

Either way you will go back to the Power Options screen.

Do you remember when we talked about the proper way to shutdown the computer? I said that you should not just push the power button. If you accidentally push the power button instead of shutting the computer down, this is where we tell Windows what to do.

Click on the "Choose what the power button does" option

The System Settings Dialog box will come to the screen. This is shown in Figure 2-11.

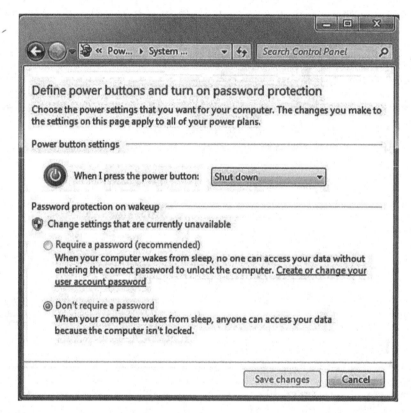

Figure 2-11

The top part allows you to change what the computer will do if you press the power button. Your choices are: Do nothing, sleep, hibernate, or Shutdown. I believe it is set to shut down by default. I can only guess that Microsoft figured that if you pressed the power button, you wanted to turn the computer off. This will not just kill the power and let the computer die. This will do an organized shut down. I still do not recommend that you shut your computer down this way. Do it the correct way.

When your computer wakes up from being idle too long, do I want a password to be entered? This will be the same password that you use to logon to the computer. If you do not use a password to logon to the computer, this setting will have no effect on your computer.

76

Lesson 2 – 5　　Themes

Themes have been a part of Windows for a while. This is a way for you to personalize your computer by having the background, window colors, sounds, and screen saver change all at one time. In this lesson we will be looking at the different themes.

Bring the Personalization screen back to the monitor

More than likely, you started out with the Windows 7 Basic theme. This is the one that has the fish on it and a blue background. Each theme has its own set of pictures that are standard with it. You can choose to use a picture from another theme as the background, but you will lose some of the available effects. We will discuss these as we continue. Figure 2-12 shows some of the available themes.

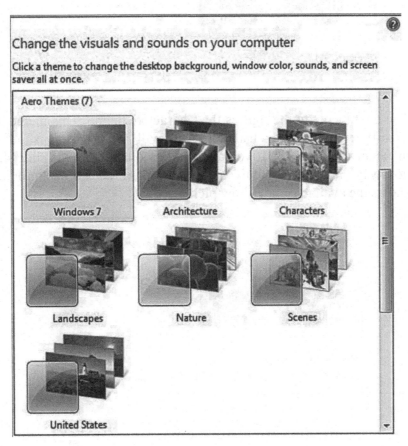

Figure 2-12

The Windows 7 Basic theme, seen at the top left, can have a single picture as the background or you can select the pictures you would like to have for a background, and have the background change every so often. You can get the picture from the suggested pictures that come with the theme or you can choose your own picture from ones that you have saved on the hard drive.

The next theme to the right is the Architectural theme. This theme has six pictures of different architectural designs that will rotate to show different backgrounds. . If you click on any one picture the slide show option will go away and you will only have one picture as the background. Figure 2-13 shows the available slide show pictures.

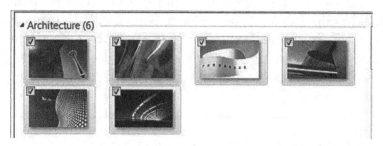

Figure 2-13

The next theme to the right is the Characters theme. This theme has six pictures of different animated characters that will rotate like a slide show to show different backgrounds. If you click on any one picture the slide show option will go away and you will only have one picture as the background. Figure 2-14 shows the available slide show pictures.

Figure 2-14

The next theme to the right is the Landscape theme. This theme has six pictures of different landscapes that will rotate like a slide show to show different backgrounds. If you click on any one picture the slide show option will go away and you will only have one picture as the background. Figure 2-15 shows the available slide show pictures.

Figure 2-15

The next theme to the right is the Nature theme. This theme has six pictures of various flowers that will rotate like a slide show to show different backgrounds. If you click on any one picture the slide show option will go away and you will only have one picture as the background. Figure 2-16 shows the available slide show pictures.

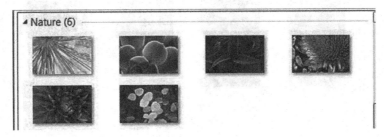

Figure 2-16

The next theme to the right is the Scenes theme. This theme has six pictures of various scenes (I am not sure what else to call them) that will rotate like a slide show to show different backgrounds. If you click on any one picture the slide show option will go away and you will only have one picture as the background. Figure 2-17 shows the available slide show pictures.

Figure 2-17

The last theme in the Windows Themes group is the Theme for United States. This theme has six pictures of various U.S. backgrounds that will rotate like a slide show to show different backgrounds. If you click on any one picture the slide show option will go away and you will only have one picture as the background. Figure 2-18 shows the available slide show pictures.

Figure 2-18

Directly below these themes are the Ease of Access themes. These themes do not have a slide show available with them and only allow one image or one solid color with them.

The themes with the slide show also have a setting that will tell Windows how often to change the background picture. This can range from every ten seconds to only once a day.

Look through the different themes and see if you can find one that fits your personality, if you do save it

Chapter three   Working with a Window

In order to understand how Windows 7 works, it is best to jump right in and start with a program. This will allow you to actually work with the program and Windows 7. We will start out by opening a program that comes with Windows 7. Next we will go through the parts of the window. After this we will look at the Maximize, Minimize, and Restore features. This will be followed by resizing and moving the window around on the screen. We will finish this chapter by closing the window and exiting the program.

Lesson 3 – 1   Starting a Program

A program is a set of instructions that tells your computer how to do something. If you were baking a new type of cake, you would get out the recipe and follow the directions given in it. In order for your computer to work, it must also have a set of instructions to follow. These instructions are very exact and complex. These instructions are called a program. The word processor on your computer is a program. Internet Explorer that you use to view websites is also a program. The game Solitaire is also a program. This lesson is designed to show you how to open (or start) a program.

Click the Start button

The start menu will pop up. If you see an icon for the desired program, click it, and Windows loads the program. For now we want a particular program that is not showing.

You may wonder why some programs appear above the All Programs button when you click the Start Button. This is the Recently Opened Programs List that Microsoft provided for quick access to the programs we have recently used.

Move your mouse to All Programs and let it hover there for a moment

A menu of all of the programs (or program categories) will pop up above the all programs button.

Click on the Accessories file

Another menu will pop out below the Accessories file. The program we want to use is the WordPad program. Figure 3-1 shows the menus.

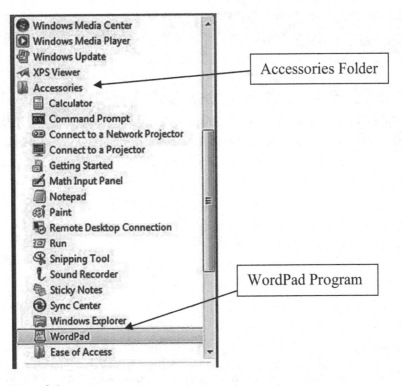

Figure 3-1

Click on the word "WordPad"

You can tell if you are in the correct spot to click the mouse because the mouse pointer will turn into a hand. Also, when the mouse pointer turns into a hand you will know that you only have to click the mouse once to open the program not twice. If the mouse pointer does not turn into a hand, like when you move it to a Desktop shortcut, you will need to double-click the mouse to open a program.

The WordPad program opens on the screen in its own window. You should also notice that a button for the program appears in the Task Bar. Figure 3-2 shows the WordPad screen. Figure 3-3 shows the same program in the Windows versions before Windows 7.

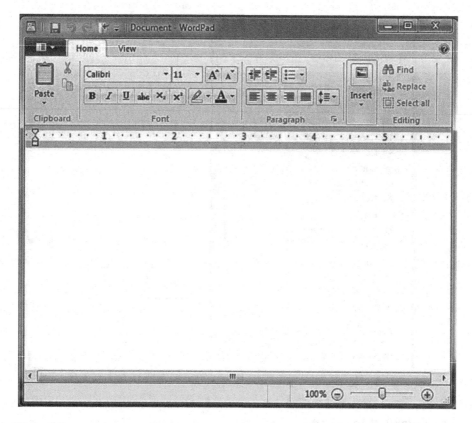

Figure 3-2

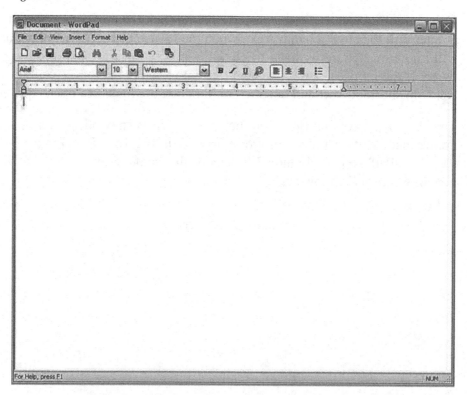

Figure 3-3

As you can see the two programs look different. They are the same program, only the version in Windows 7 has a completely different look and feel. This new version looks more complicated, but I can assure you that it is easier to use, once you get past the initial shock.

That is all there is to opening a program, and this was the hard way. If the icon for the desired program was on the Start Menu as a Recently used program or Pinned Item, all you would have to do is click on the icon. Let's not forget that you may have made a shortcut to the program on the Desktop or Taskbar. If it was on the Desktop, you would have to double-click on the icon to open the program. If it was on the Taskbar you would only have to click on the icon.

Lesson 3 – 2   The Parts of a Window

Now that you have seen two different versions of the same program, you will need some explanations on the new version. The first thing that you probably noticed was that the top part of the program looked very different. The older version had the menu across the top that you have seen in Windows programs since the very early days. The new version has the Ribbon. The older version had toolbars that you could click on for easy access to some of the more common features. The new version has the Quick Access toolbar that you can customize. Now that all of that is out of the way, let's look at the WordPad screen. Figure 3-4 shows the WordPad screen.

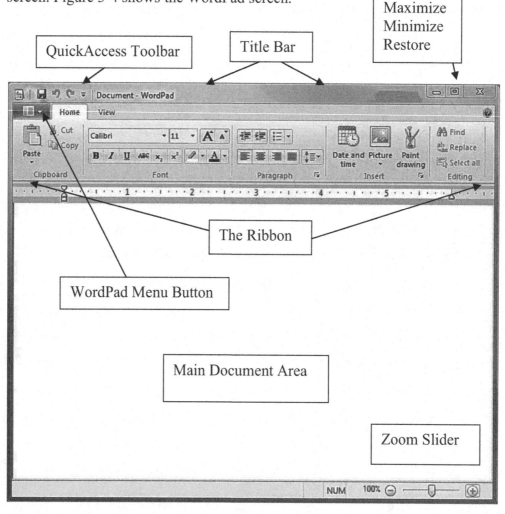

Figure 3-4

I will go through the various parts of the program screen and try to give you a definition and the function of this each part of the screen.

Title Bar:

The Title Bar is located at the top of the window and the name of the program is displayed on it. Also displayed on the Title Bar is the name of the document that is being displayed. Figure 3-5 shows the Title Bar.

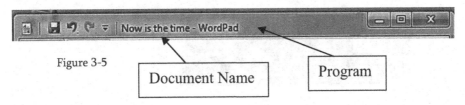

Figure 3-5

Document Name

Program

Minimize Button:

The Minimize button is located on the right end of the Title Bar. There are three buttons on the right end of the Title Bar. The one on the left (the one with the small line at the bottom) is the Minimize button. Clicking the Minimize button will cause the program to shrink down until it is no longer visible on the screen. The only thing visible will be the button on the Taskbar. You will have to click on the program icon in the Taskbar to get the program back on the screen. Figure 3-6 shows the Minimize button.

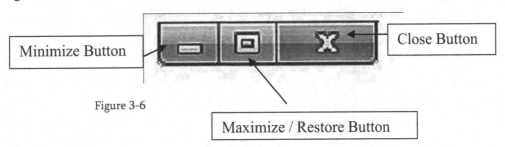

Minimize Button

Close Button

Figure 3-6

Maximize / Restore Button

Maximize / Restore Button:

The Maximize / Restore button, located next to the minimize button, has two functions. If the program window does not fill the entire monitor screen, clicking this button will cause the program window to maximize and fill the entire monitor screen. If the program window has already been maximized the function of the button will change. If you click the button it will cause the program window to go back to the size it was before it was maximized.

Close Button:

The Close button, located at the very end of the Title Bar, will close the program when clicked. If you have made any changes to the program that is running, such as adding or deleting text if it was a word document, you will be prompted to save any changes that were made.

The Quick Access Toolbar:

The Ribbon, as you will find out, is wonderful, but what if you want some commands to always be right at your fingertips without having to go from one tab to another? Microsoft gave us a toolbar for just that purpose. This toolbar is called the Quick Access Toolbar and is located just above, or below, and to the left end of the Ribbon. The Quick Access Toolbar is shown in Figure 3-7. The Quick Access Toolbar will be discussed in detail in Chapter four.

Figure 3-7

The WordPad Menu Button:

The WordPad Menu Button, located just below the Quick Access toolbar, is very similar to the old "File" button in the previous versions of WordPad. By clicking this button you can create a new document, open an existing document, save a document, print a document, change the settings on the page, send the current document as an email, or exit the program. Figure 3-8 shows the drop down menu for the Menu button.

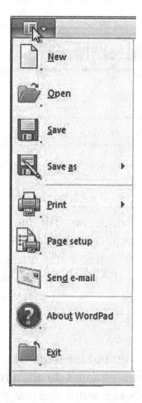

Figure 3-8

The Ribbon:

The Ribbon is something new to WordPad and most of the other programs in Windows 7. This was first made available in Microsoft Office 2007 and is now standard on most of the programs in Windows 7. The Ribbon is what you will use to access the clipboard, format text and paragraphs, and insert objects into your document. The Ribbon will be discussed in detail in Chapter four.

Main Document area:

This is where you will enter the text for your document and insert any objects, such as pictures. Again this will be covered in Chapter four.

Zoom:

The zoom feature will allow you to zoom into the document to make the text appear larger than it is for easier viewing on the screen.

Lesson 3 – 3   Minimize, Maximize, and Restore

In the last lesson we mentioned the Minimize, Maximize and Restore buttons. In this lesson we will try these buttons out and make sure you understand how they work.

Open the WordPad program if it is not open

The WordPad program will open and may or may not fill the entire screen. For the first part of the lesson it will not matter if it fills the screen or not. We are going to "Minimize" the window first. Figure 3-9 shows the buttons again.

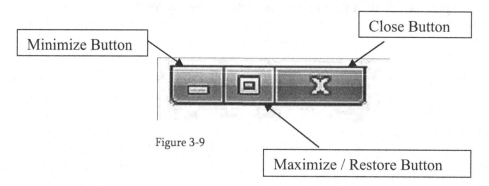

Minimize Button

Close Button

Maximize / Restore Button

Figure 3-9

I will repeat what I said before. Clicking the Minimize button will cause the program to shrink down until it is no longer visible on the screen. The only thing visible will be the button on the Taskbar.

Click the Minimize button

The WordPad program will completely disappear from the screen. The only reference to the program should be the button on the Taskbar that has a small icon of the program on it (See Figure 3-10). To bring the program back to the screen we have to click the icon on the Taskbar.

Figure 3-10

Click the WordPad button on the Taskbar

The program will jump back onto the screen, ready for you to work on it.

Next we will look at the Maximize / Restore button. Figure 3-9 showed you the Maximize button and Figure 3-11 shows the Restore button.

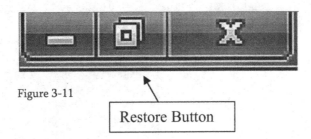

Figure 3-11

Restore Button

As you can see the Restore button is different than the Maximize button.

Click the center Maximize / Restore button

If your window filled the screen, it will shrink back to a smaller size. If it did not fill the screen, it will grow until it does fill the screen.

Click the button again until you are sure that you have seen both views

In the off chance that the screen did not seem to change, this will be covered in Lesson four.

Close the WordPad program by clicking the Close button and if it asks if you want to save any changes made, click the Don't Save button

We closed this program so that we could look at the screen of another program.

Here is another program that comes with Windows 7 and also utilizes the Ribbon.

The paint program is a drawing program that will let you draw and edit pictures. It is located in the Accessories folder just like the WordPad program.

Click the Start button and then move the mouse to All Programs

Click on the Accessories folder and then click on the Paint program

The Paint program will immediately, well almost immediately, jump onto your screen. The Paint program screen is shown in Figure 3-12.

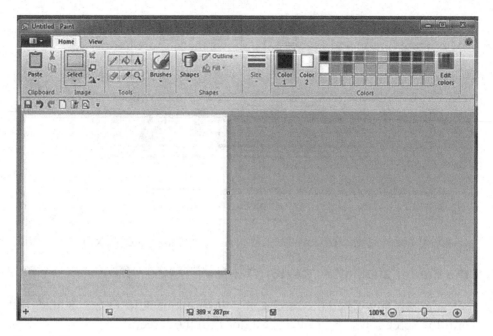

Figure 3-12

As you can see the paint program also utilizes the Ribbon, instead of the menus and Toolbars that the older versions used.

You may also notice some similarities and differences between the two Ribbons. Both Ribbons have a "Home" Tab, a "Menu" button, and a "View Tab. The Groups and commands on the tabs are not the same for both programs. Each program has different commands that you will use to work with the programs. We will work more with the Ribbon a little later and this will probably make more sense then.

Close the Paint program

Lesson 3 –4   Moving & Resizing a Window

In lesson 3-3 we discussed using the Minimize and Maximize buttons to change the size of the window on the screen. What if you want the window size to be somewhere in between the two extremes? Sometimes you may want to move a window to a different location on the screen, without filling the entire screen with the window.

This lesson will show you how to move the window to a new location on the screen and change the size of the window.

First make sure the WordPad program is running and does not fill the entire screen. If the window is maximized and fills the entire screen you cannot move the window.

Open the WordPad program if it is not open

If the program window is maximized (it fills the entire screen), click the Restore button.

You can move the entire window around on the screen by clicking the Title bar and dragging the window anywhere on the screen.

Move your mouse pointer to the Title Bar, when it is there click and hold the left mouse button down

While you are still holding the left mouse button down drag the mouse across the screen

As you drag the mouse, the window will also move on the screen.

When you have the window in the position you want, release the mouse button

The window will now stop moving and be in the desired location.

If the window is generally where you want it, but it is either too large or too small, you can change the size of the window.

At the very bottom of the window and on the right side you should see a series of small lines forming a triangle (See Figure 3-13).

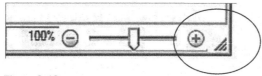

Figure 3-13

If you move your mouse to this part of the screen, you can change the size of the window. When you move your mouse over these lines, the pointer will change into a white double sided arrow and it will be at an angle. It will look similar to this (↖).

Move the mouse pointer to the lower right side of the window

When it changes into the double sided arrow, click and hold the left mouse button down

Move the mouse up and down and then to the right and left and watch the window size change

When you are satisfied with the window size, release the mouse button.

Using this procedure will allow you to change the height and the width from one location. You can, however, make the adjustment from a different location on the screen and only change the height or the width.

Position your mouse pointer over the right border of the WordPad program as shown in Figure 3-14 (The arrow shown in the figure is larger than the one will be on your screen)

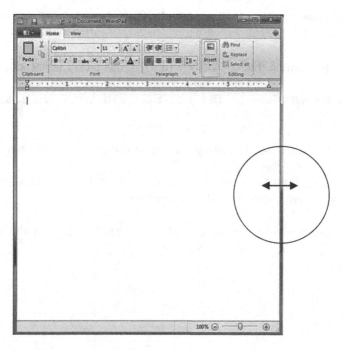

Figure 3-14

The two arrows point to the direction that you can drag the window's border. This allows you to change the size of the window to suit your specifications. Using the above situation, you could move the border to the left or to the right. If you had positioned the pointer on the top or the bottom, you could move the border either up or down. If you had positioned the pointer on one of the corners, you could move both the vertical and horizontal borders at one time.

Click and hold down the left mouse button and drag the mouse to the right about one inch.

Notice how the window stretches as you move the mouse.

Release the mouse button.

The window is now displayed in its new size.

Repeat the above process only this time position the mouse pointer over the bottom border of the window

Moving the mouse will cause the window to be either shorter or taller depending on which way you move the mouse.

Lesson 3 – 5   Closing a Window

When you are finished working with a program or window, you can close it to remove it from the screen and the computer's memory. You can close any window by clicking its Close Button which appears in the upper right corner.

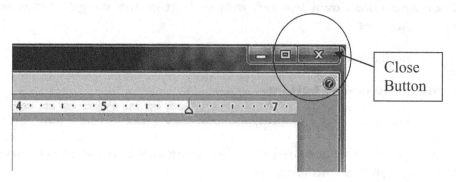

Figure 3-15

Move the pointer over to the close button and click the left mouse button.

The WordPad program closes and is no longer visible on the screen. You will also want to notice that the button for the program is no longer in the Taskbar. At this point it is advisable to inform you that in order to save anything you have typed into the word document you must save the changes before you close the program. Since we have not typed anything, it is okay to close the program.

If the window happens to be minimized when you want to close it, you will need to restore it by clicking on its name on the Task Bar before you click its close button. Of course you could always right-click the icon on the Taskbar and choose close from the shortcut menu.

That is all there is to closing a program or window.

Chapter Four   Working with a Program

For this chapter, we are going to switch gears. In the previous chapters we were working with the Windows 7 operating system, now we will be working with a Windows program. We will learn how to control a program using the Ribbon and Quick Access Toolbar, as well as how to fill out a dialog box. Since every program is a little different, I will not be able to cover every possible Ribbon and Tab that you might see. The good news is that once you learn how the Ribbon and Quick Access Toolbar works in one program, the procedure for using them will remain the same for all programs.

We will have the opportunity to learn the most by using a program that comes standard in Windows 7, WordPad. WordPad will give us the chance to use the Ribbon as well as the Quick Access Toolbar. We will be able to fill out a dialog box and enter, edit, and delete text. You will be able to take the skills you learn in WordPad and apply them to other Windows programs.

Lesson 4 – 1   The Ribbon an Overview

The Ribbon is new to most of the programs in Windows 7. Microsoft first released the Ribbon in Office 2007 and in my opinion it transformed the way a person uses the Office series of products.

The Ribbon has been designed to offer an easy access to the commands that you (the user) use most often. You no longer have to search for a command embedded in a series of menus and submenus. The Ribbon has a series of Tabs and each tab is divided into several groups of related commands. Figure 4-1 shows the Ribbon across the top of the WordPad program.

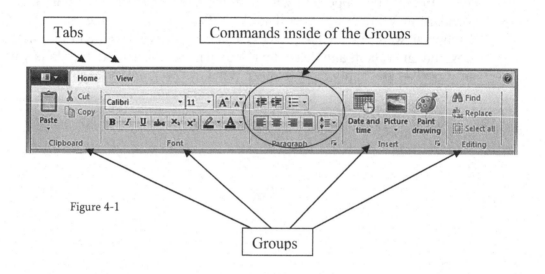

Figure 4-1

There are three major components to the Ribbon.

Tabs:

There are two basic tabs across the top.

> The Home Tab contains the commands that you use most often.

> The View Tab allows you to change to the different views that are available.

Groups:

Each Tab has several Groups that show related item together.

Look at the Home Tab to see an example of the related Groups.

The Home Tab has the following Groups: Clipboard, Font, Paragraph, Insert, and Editing.

Commands:

A Command is a button, a box to enter information, or a menu.

The Clipboard Group, for example, has the following commands in it: Cut, Copy, and Paste.

When you first glance at a group, you may not see a command that was available from the menus of the previous versions of WordPad. If this is the case you need not worry. Some Groups have a small box with an arrow in the lower right side of the Group. See figure 4-2 for a view of a group with this arrow.

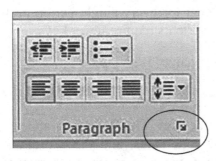

Figure 4-2

This small arrow is called the Dialog Box Launcher. If you click on it, you will see more options related to that Group. These options will usually appear in the form of a Dialog Box. You will probably recognize the dialog box from previous versions of WordPad. These options may also appear in the form of a task pane. Figure 4-3 shows the Paragraph Dialog Box.

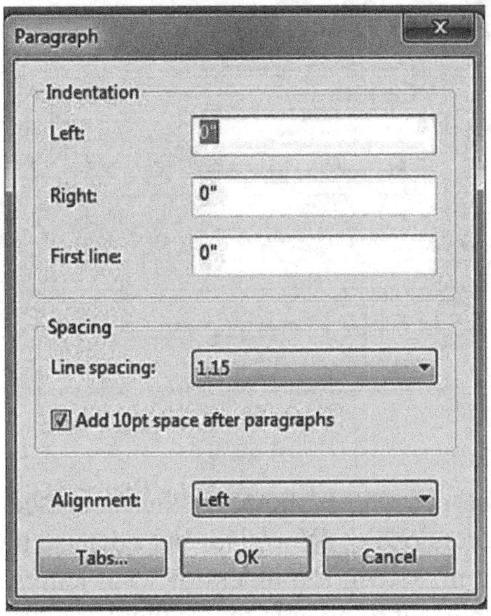

Figure 4-3

Speaking of previous versions, if you are wondering whether you can get the look and feel of the older versions of WordPad back, the answer is simple, **no you can't**.

The good news is that after playing with and using the Ribbon, you will probably like it even better. It really does make working with the word processor easier. The Ribbon will be used extensively and the tabs will be covered in more detail later as we go through this book.

Are you ready for even more good news? Microsoft has included new shortcuts with the Ribbon. Why you might ask. It is because this change brings two major advantages. First there are shortcuts for every single button on the Ribbon and second because the many of the shortcuts require fewer keys.

The new shortcuts also have a new name: **Key Tips**

Press the ALT key on the keyboard

Pressing the Alt key will cause the **Key Tip Badges** to appear for all Ribbon tabs, the Quick Access Toolbar commands, and the WordPad Menu Button. After the Key Tip Badges appear, you can press the corresponding letter or number on the badge for the tab or the command you want to use. As an example, if you pressed Alt and then H you would bring the Home tab to the front. Figure 4-4 shows what the Ribbon looks like after pressing the Alt key.

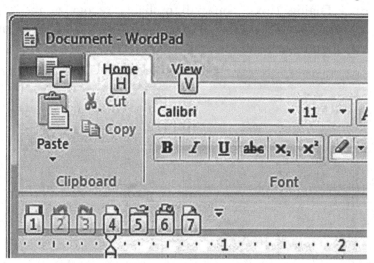

Figure 4-4

Note: You can still use the old Alt + shortcuts that accessed the menus and commands in the previous versions of WordPad, but because the old menus are not available, you will have no screen reminders of what letters to press, so you will need to know the full shortcut to be able to use them.

Lesson 4 – 2   Quick Access Toolbar

The Ribbon, as you will find out, is wonderful, but what if you want some commands to always be right at your fingertips without having to go from one tab to another? Microsoft gave us a toolbar for just that purpose. This toolbar is called the Quick Access Toolbar and is located just above, or below, and to the left end of the Ribbon. The Quick Access Toolbar is shown in Figure 4-5.

Figure 4-5

The Quick Access Toolbar contains such things as the Save button, the Undo and Redo buttons. These are things that you normally use over and over and you will want them available all of the time.

There is even more good news, if you want to add an item to the toolbar, the process is very simple. At the right end of the toolbar is an arrow pointing downwards. If you click on this arrow, a new drop down menu will come onto the screen, as shown in Figure 4-6.

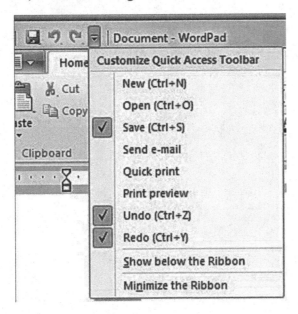

Figure 4-6

From this menu you can choose to add any of the available choices. If you are going to use this program very often, I would consider adding the following shortcuts to the Quick Access Toolbar: New, Open, Quick Print, and Print Preview. These are commands that you will constantly use.

To add an item to the toolbar all you have to do is click on it. Each time you click on an item the menu will disappear and the item will be added to the toolbar.

Add the following items to the Quick Access Toolbar:

New

Open

Quick Print

Print Preview

You can also choose to show the tool bar below the Ribbon instead of above it. I have my computer set to show the Quick Access Toolbar below the Ribbon, probably because the toolbars were always below the menu bar in the older versions.

As we continue in this book, you will find out that the Quick Access Toolbar is your friend and you will use it a lot.

Lesson 4 – 3   The Ribbon: A Closer Look

In lesson 4-1, we had an overview of the Ribbon. In this lesson we will take a closer look at the different parts of the Ribbon.

In this lesson we will look at each tab of the Ribbon and each group that is on the Tab.

Open the WordPad program if it is not open

We will start our discussion of the Ribbon with the Home Tab. This should be in the front when we open WordPad.

Starting on the far right is the Clipboard Group. Figure 4-7 shows the Clipboard Group.

Figure 4-7

This group deals with cutting, copying, and pasting selected text and objects. This is one of the most used groups in WordPad.

The next group is the Font Group and is shown in Figure 4-8.

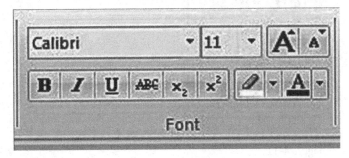

Figure 4-8

As you would expect, this is where you would perform all of the formatting for the text.

The next group is the Paragraph Group. This group is shown in Figure 4-9.

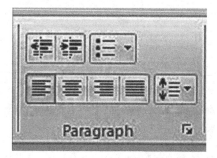

Figure 4-9

This group contains the formatting for the paragraphs. This includes the text alignment as well as the line spacing and inserting lists into your document.

The next group is the Insert Group and is shown in Figure 4-10.

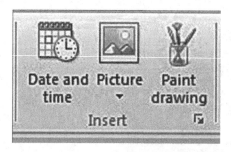

Figure 4-10

This group deals with inserting objects into your document. This could be the date and time, a picture on your hard drive, or a drawing that you have painted using the Paint program.

The last group on the Home Tab is the Editing Group. This is shown in Figure 4-11.

Figure 4-11

This group deals with searching for and replacing text.

Click your mouse on the View Tab

The View Tab, shown in Figure 4-12, has three groups on it.

Figure 4-12

The first group on the View Tab is the Zoom Group and is shown in Figure 4-13.

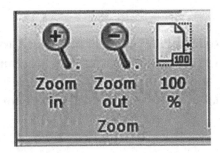

Figure 4-13

This group will allow you to view the text in either a larger or smaller size to make the viewing easier.

The next group is the Show / Hide Group. This is shown in Figure 4-14.

Figure 4-14

This group will allow you to show the ruler and/or Status bar on the screen.

The last group is the Settings Group and is shown in Figure 4-15.

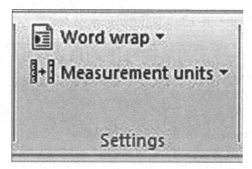

Figure 4-15

This group will allow you to change the settings for the Word Wrap feature and the settings that are used for measuring.

The Ribbon may seem confusing to you right now, but I assure you that you will come to love the Ribbon. It is so much easier than searching through the menus and submenus trying to find a command. Well perhaps I can't assure you that you will love the Ribbon, but I know that I love it and I am not easy to please.

Lesson 4 – 4   Dialog Boxes

There are times when the Ribbon does not hold every command that is available to you. Sometimes there are just too many things that you can do. An example of this is the Insert Group. There are other commands that can be performed and there is not enough room on the Ribbon to show them all. This is one of reasons Microsoft created the Dialog Box. The other reason is that the dialog box can show all of the commands available and allow you to make more precise adjustments.

Open the WordPad program if it is not open

Click on the Home Tab to make sure that this tab is on the front and showing

The Paragraph Group, shown in Figure 4-16 has a dialog box associated with it.

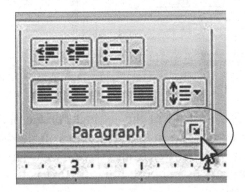

Figure 4-16

There is a small square in the lower right corner that has an arrow on it. This is the Dialog Box Launcher. You launch the dialog box by clicking on this small square.

Move your mouse over to the dialog box launcher

When your mouse gets over the dialog box launcher, it will turn into an amber color. This is how you will know that you are ready to click the mouse.

When the mouse is in position, click the left mouse button

The Paragraph Dialog Box will come onto the screen (See Figure 4-17).

107

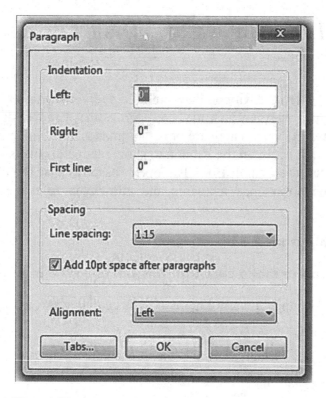

Figure 4-17

From this dialog box you can adjust all of the settings for the paragraph at one place. You can set the exact amount of indentation that you need. You can also set the spacing between each line and determine if you want any extra spacing between the paragraphs. You can also choose how the text is to be aligned. All of this can be done from one place.

Click the Cancel button to close the dialog box

Dialog boxes can be time saving and they are convenient.

Lesson 4 – 5 Creating a Document

Okay, you have done a lot of reading and very little work. Let's create a document and use some of the things we have been talking about. First we will create a document so that we have something to work with and then we will start using the Ribbon.

If necessary, open the WordPad program

When you open the program you will have a blank document on the screen.

Type the following memo

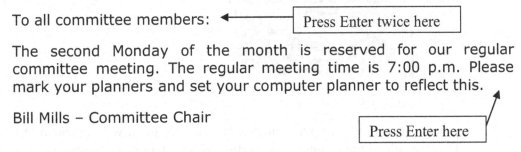

That is all that there is to creating a document. That was pretty easy, wasn't it?

You will notice that I had you press Enter in a couple of places but not in other places. Every time we press the Enter key, a new paragraph is started. We pressed Enter twice to ensure that there would be a blank line between the first line and the main body of our memo. We did not press the Enter key during the main body of the memo because the text will automatically expand to the next line. This is called Word Wrap. The text will automatically wrap to the next line.

Note: A new document can be created at any time. There is, of course, more than one way to create a new document. One way, and it seems this is the most common way, is to click the WordPad Menu Button and select "New" from the drop down list. An easier way is to click on the "New" command that is in the Quick Access Toolbar (provided that you added it to the toolbar). You can also use a keyboard shortcut to create a new document. This is done by pressing the ALT key and then pressing the "F" key and then pressing the "N" key. If you remember from the earlier days, this was the same keystrokes that were used in the older version of WordPad that had the menu bar across the top.

Regardless of the method you use to create a new document, you will have to make a second choice as to the type of document you want to create. For the most part you will choose to create a Rich Text Document.

Lesson 4 – 6   Saving a Document

Now that you have gone through all of the trouble of entering the text, we need to save it so that nothing will happen to our memo. I don't know about you, but I have gone through situations like this only to have the power go off and everything is gone. Before that happens, let's learn how to save the Document.

Click the WordPad Menu Button

The WordPad Menu Button is located directly to the left of the Home Tab and is shown in Figure 4-18.

Figure 4-18

When you click the Menu Button a drop down menu will appear on the screen. This lists all of the things you can do with the document. Figure 4-19 shows the drop down menu.

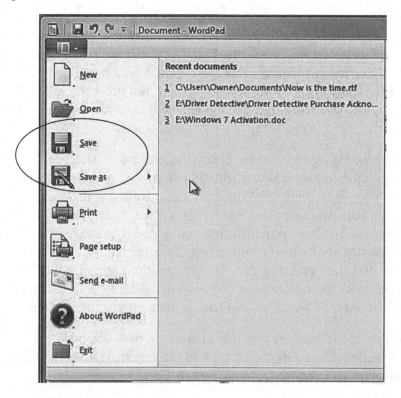

Figure 4-19

There are several things that you can do from here. You can start a new document, open an existing document, save a document, print a document, etc. I want you to notice that there are two save commands: a "Save" and a "Save As" command.

If this is the first time you have saved the current document, The Save choice will bring the Save As dialog box to the screen. If this is not the first time you have saved the document the choice will have different results. Now would be a good time to explain the difference between the two of them.

Save: This will replace the existing document with the newer version that is displayed on your screen. All changes made will be saved and the original document, before you made any changes, will be gone.

Save As: This will allow you to save the currently displayed document with a different name. This will allow you to keep the original document just as it was before any changes were made to it and the new version will also be saved only under a different name.

Move the mouse to the Save As choice

Another menu will slide out to the side.

The first thing you have to do is make an important decision: what format do I want to save this document in. Figure 4-20 shows the choices you have to choose from.

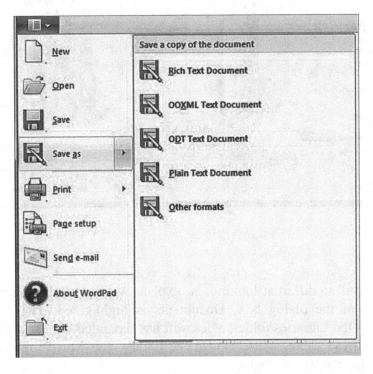

Figure 4-20

Of all the different formats available, there are only two that might interest you. The first is the default format which is the Rich Text Document. The other is the Plain text document.

The "odt" and "ooxml" formats are compatible with Open Office which is normally a free Office package available online. I have used the Open Office that is available and it does not compare, in my humble opinion, with the Microsoft Office package.

The plain text document removes all formatting and saves the document as plain text. This is not normally what you will want. The default, the Rich Text Document, is the format you will want to use.

Click on Rich Text Document

Something new has popped up on the screen. This is the new look for the Save As Dialog box, and is shown in Figure 4-21.

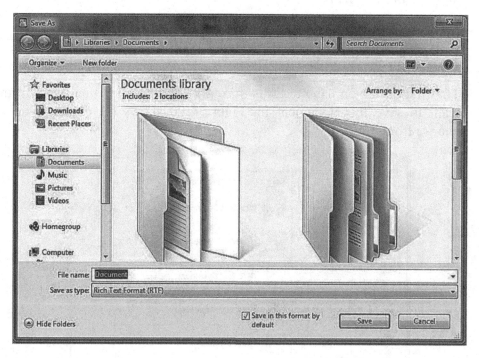

Figure 4-21

Since this is so different looking, an explanation is probably necessary. On the left side of the dialog box, Documents is highlighted. The Documents are located in the Libraries folder. Microsoft has separated the various things we use into libraries. There is the Document library, the Music library, the picture library, and the Video library.

We will want to keep our documents in the Document library. But that is not all there is to it. Inside the Documents library are two folders.

Click the small triangle next to the Documents icon

Figure 4-22 shows the triangle if you are having trouble finding it. If it is not visible, move your mouse over the "Documents" name and the triangle will become visible.

Figure 4-22

These are the "My Documents" folder and the "Public Documents" folder. Any document that we want to share with other users on this computer should be saved in the Public Documents folder. That makes sense; if these are not private and you plan on sharing them with all of the other users on the computer (or network), why not put them in the Public Documents folder. If, however, these are your personal documents, you would want to save them in the My Documents folder so no one else can see them.

Click on the My Documents **folder**

If there is a small triangle next to it, do not click on the triangle at this time. We will discuss this more in the next chapter.

Click in the textbox next to File Name in the bottom section

This is where we will give our document a unique name. Every document must have a unique name to identify it and keep it separate from all other documents.

When you click the mouse inside of the textbox the default name of "Document" will be highlighted. When we start typing, the default name will be replaced by whatever we type.

Type the name Committee Memo **in the textbox**

When you are finished typing, click the Save button

The dialog box will disappear from the screen and the document will be saved as a WordPad file called Committee Memo. You will be able to access this file any time you want by locating it under My Documents and double clicking it with the mouse. We will discuss opening an existing document a little later.

That is all there is to saving a file. Remember use Save As if you wish to keep the original document as it was before any changes were made and use Save if you wish to replace the original file with the revised version.

Note: From now on, when you are asked to save a document, make sure you save it in the MY Documents folder until you are told differently.

I know that it sounded like a lot to go through, but the document is now permanently saved on your hard drive.

Now that we have saved our work we can safely close the file.

Closing a file in WordPad is different than closing a file in Microsoft Office Word. In Word there is a feature that will let you close the file and still leave the program open to start a new document or open another file. That is not the case with WordPad.

When you click the WordPad Menu Button, you can open a different file, start a new document, or exit the program to close an existing file.

Click on the Exit command

This will cause the program to quit and close the open file.

Lesson 4 – 7   Opening a Document

Open the WordPad program

Remember that WordPad is located in the Accessories folder under all programs which is under the Start button.

There are three basic ways to open an existing WordPad Document. We will go through each method in this lesson. We will look at what is probably the most common way first.

Click the WordPad Menu Button

You have seen this before and it is shown again in Figure 4-23.

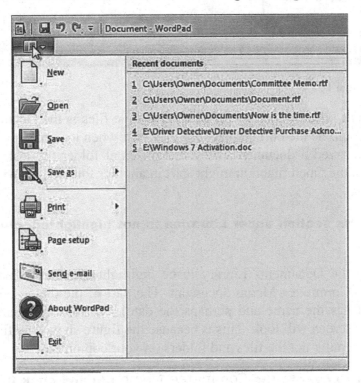

Figure 4-23

Click on the Open choice

This will bring the Open Dialog Box to the screen as shown in Figure 4-24.

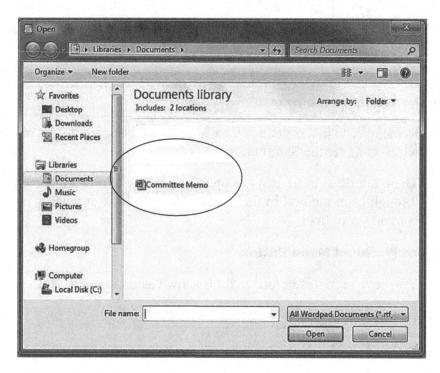

Figure 4-24

The default location that WordPad uses to save files is the Documents Library, so that would be the normal place for it to start when it opens a file. If, however, you have saved a document (file) in a different folder the last time you used WordPad, the Open function might start in another folder. This will be covered a little later.

If the Documents section under Libraries is not highlighted, click on it with the mouse

We want the Documents Library to be highlighted because that is where we saved the Committee Memo document. The part of the dialog box on the right (where it has the name and perhaps the date) will look different in the figure than your screen will look. This is because the figure shows the files and folders on my computer not the files and folders on your computer.

Move your mouse over to the Committee Memo **file and click the left mouse button**

As soon as you click the mouse on the file name the open button will change. The right side of the button will now have a downward pointing arrow on it. This drop down list will allow you to open the file or show previous versions. That deserves an explanation.

Previous versions are copies of files and folders that Windows automatically saves as part of a restore point. You can use previous versions to restore files and folders that you accidentally modified or deleted, or that were damaged.

Depending on the type of file or folder, you can open, save to a different location, or restore a previous version.

This choice is one of the new features of Windows 7.

Move your mouse over to the Open button and click on it to open the document.

If you happen to click on the down arrow instead of the actual button, you will need to click the Open option. The memo we typed earlier will open and be on the screen.

As I said before there is more than one way to open an existing document. We opened the Committee Memo the hard way first. This way is a little easier.

Instead of clicking the WordPad Menu Button and selecting the Open choice to bring the Open Dialog Box to the screen, you could have clicked the mouse on the Open command that we installed on the Quick Access Toolbar. This would also bring the Open Dialog Box to the screen. Figure 4-25 shows the Quick Access Toolbar.

Figure 4-25

There is an even easier way to open an existing document, if you have had the document opened recently.

You could click on the WordPad Menu Button

This will bring the drop down menu to the screen and is shown in Figure 4-26.

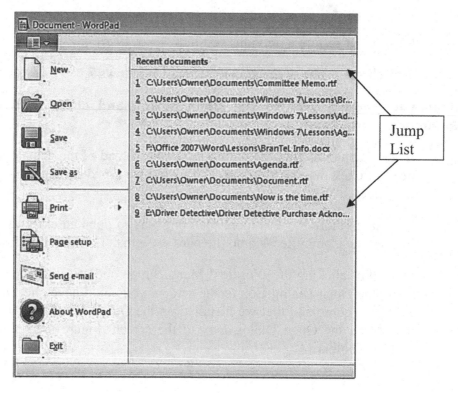

Figure 4-26

Then you would move your move to the Committee Memo name on the right side and click on it. The right side shows recently opened documents. This list of documents is called a Jump List, because it allows you to jump directly to a document.

A Jump List is also available if you right click on a program that is shown on the Task Bar. Also if the program shows up directly on the Start Menu, there is usually a Jump List attached to it. You can tell the ones with a Jump List by the arrow pointing to the right from the name of the program.

If you like using the keyboard shortcuts you could use one of the two methods described next.

Pressing and holding the CTRL button down and then pressing the O key and releasing both keys will bring the Open Dialog box to the screen. And do not forget the old ALT + F + O shortcut. That will also bring the Open Dialog box to the screen.

Lesson 4 – 8   Inserting Objects

Inserting an object can cover a lot of territory. On object might be a picture, a chart, or a table. It could also be something as simple as the time and date. In this lesson we will use the Insert Group to insert the time and date into our document.

Open the Committee Memo **document**

The insertion point (the flashing vertical line) should be at the very beginning of the document. If it is not at the beginning, click the mouse before the letter "T" at the very beginning. If you prefer using the keyboard, you can press and hold the CTRL and then press the Home key and release both keys.

Press the Enter key to create a blank line at the beginning

Press the Up Arrow on the keyboard one time

This will move the insertion point to the blank line.

If the Home Tab is not the front tab on the Ribbon, click the mouse on it

The Insert Group is toward the right end of the Home Tab. One of the commands in the Insert Group is Date and Time. This is where we have to click the mouse to insert the date and time.

Click where it says Date and time in the Insert Group

A dialog box will come to the screen and is shown in Figure 4-27.

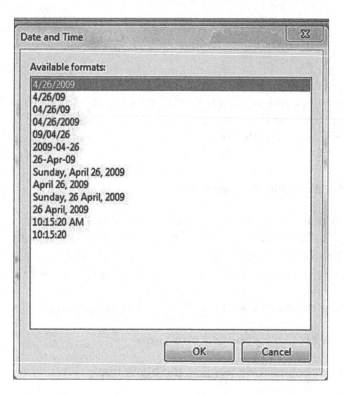

Figure 4-27

Find the choice that shows the day of the week and the date and click on it

Click the OK button

This will insert the date onto the first line of the document. Inserting a picture is just as easy as inserting the date. You just have to find the picture you want and click on it and then click the Open button.

Do not save your changes and close the file

Lesson 4 – 9   Editing Text

One of the most common things you will do on your computer is making changes to what you have typed. It doesn't matter if you are using a word processor program or send an email, more than likely you will think of a better way of communicating your thoughts. In this lesson we will make some changes to our Committee Memo.

Open the Committee Memo **document with WordPad**

When it comes to editing text in a document, the first thing you have to understand is how to select text. Before you can do anything with text you must first select it. This lesson will show you how to select text.

There are several ways to select text, so we will start with the most common.

The most common method of selecting text is by moving the mouse pointer to the text, clicking and holding down the left mouse button and dragging the mouse over the text, and then releasing the mouse button. Let's give it a try.

Select the text "To all committee members:"

To select the text, move the mouse pointer until it is just before the T in "To all committee members". Click and hold the left mouse button down and drag the mouse to the right until the rest of the line is highlighted. Now release the left mouse button.

This can be a little tricky, so you may have to try it a few times before it works.

If you did this correctly the document should look like Figure 4-28.

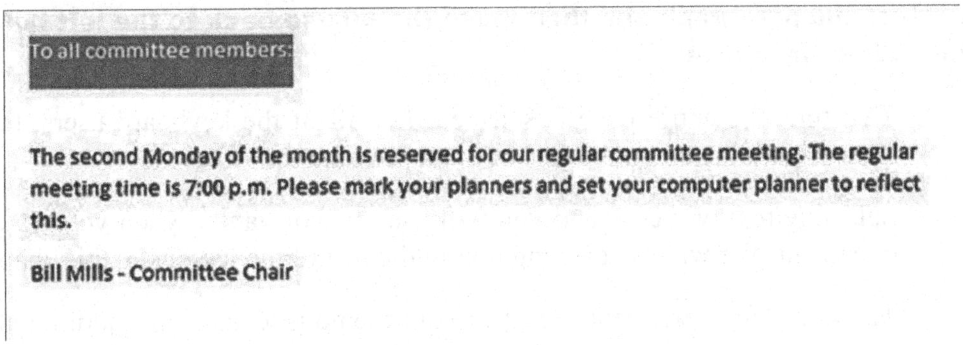

To all committee members:

The second Monday of the month is reserved for our regular committee meeting. The regular meeting time is 7:00 p.m. Please mark your planners and set your computer planner to reflect this.

Bill Mills - Committee Chair

Figure 4-28

As you can see, when the text is selected it shows up as being highlighted.

Text can be de-selected by moving the mouse anywhere and then clicking and releasing the left mouse button.

This is not the only way text can be selected. Text can also be selected by clicking on it.

Click the mouse on the word Monday **in the first paragraph**

The insertion point will just sit there flashing inside the word. This is just what it is suppose to do when you click the mouse once.

With the mouse pointer inside the word, double-click the mouse

Monday should now be highlighted.

Click anywhere outside of the highlighted word to unselect it

Now you know two ways to select text, but these are not the only way you can select text.

Move the mouse to the left margin of the document next to the first line of the first paragraph until it changes into a small white arrow

Click the mouse once

The entire line is highlighted. This can be handy if you need to make changes to the line, such as deleting or moving the entire line.

Click somewhere off the line to unselect it

Move the mouse back to the left border and double-click the mouse

This time the entire paragraph is highlighted.

Unselect the paragraph and then move the mouse back to the left border and triple-click the mouse

This time the entire page is selected. For all of the keyboard users, this is the same as holding down the Ctrl key and pressing the A key.

Selecting text was easy and you will use this constantly when you are editing a document. We will be selecting text in the rest of the lessons in this chapter.

The first change we want to make to our memo is to add some additional text.

Click your mouse immediately after the period at the end of the first paragraph and press Enter.

This will start a new paragraph.

Type the following and press Enter when you are finished.

The July meeting will not be on this date, as it is my birthday. The meeting for July will be held on the third Monday.

That is all there is to inserting text. It really is that easy. Let's try it again.

Click your mouse at the end of the line Bill Mills – Committee Chair **and press Enter.**

This will start another new paragraph.

Type the following and press Enter when you are finished

P.S. Presents for the chair on his birthday will not be considered as sucking-up, just common courtesy.

See it really is that easy.

Save the changes that you have made

If we are going to talk about editing text we need to consider deleting text from the document.

There will be times when you decide that something you put into a document should not be there and you will need to delete it. WordPad has made this so easy you will not lose any sleep over how to get this done.

Select the last paragraph of the memo

Let's try a new way to select the paragraph.

Click the mouse on the word birthday in the paragraph.

This will move the insertion point inside the paragraph. Now comes the fun part. If you double click the mouse, you will select the word birthday, but if you triple click the mouse, you will select the entire paragraph (you can also select the entire paragraph by double-clicking in the left margin of the paragraph).

Triple click the mouse

This will select the entire paragraph.

On the keyboard, press the Delete key

The entire paragraph is now gone. Deleting and inserting text are some of the easiest things you will do in WordPad.

Close the document without saving the changes

Lesson 4 – 10   Undo & Redo

Before we continue on in our discussion of editing, there is one other thing that will come into play: Using the Undo and Redo feature.

Being human, sometimes we make mistakes. If I had not been forced to use this feature over and over again during the writing of this book, it probably would not seem as important as it actually is. Microsoft, in their infinite wisdom, looked into the future and knew that I was going to use their product and added this feature probably just for me. I will explain it to you just in the off chance you may need it.

Guess what? Undo is not on the Ribbon. Just when you thought the Ribbon held everything you would ever need, it doesn't have the Undo button.

The Quick Access Toolbar is shown in Figure 4-29, and is the home of the Undo and Redo buttons.

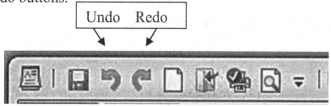

Figure 4-29

Any time you copy, type, cut, paste, or do almost anything the Undo button becomes available. If you make a mistake, you can click the Undo button and everything will be as it was before you made the mistake. Now let's be realistic here, Windows 7 will not know that you didn't really want to do the silly thing that you just did. That means that you can't make the mistake today and tomorrow when you realize that you made the mistake, expect Windows to undo it. If you click the Undo button, Windows will undo the last thing that you did, not the mistake you made five minutes ago. Obviously this may not be magic, but it is close.

Open the Committee Memo **document**

Select the line To: All Committee Members **and press the Delete key**

This line is now removed from the document, it disappeared. Oh No, you didn't really want to delete that line! We can have Undo correct the mistake.

Click the Undo button on the Quick Access Toolbar

The line "To: All Committee Members" is now back in the document.

Suppose that after careful consideration you really did want to delete the line. Do we have to re-select it and delete it again? No, Microsoft wouldn't do that to us. There is another button that will let us redo the delete. This button is right next to the Undo button and is called the Redo button.

Click the Redo button

The line is now gone again. As I said it may not be magic, but it is close.

Save your changes and close the document

Lesson 4 – 11   Formatting Text

WordPad allows you to put emphasis on text in your document by making the text darker and slightly heavier. This is called Bold. You can also make the text slanted (italics), or make the text larger (or smaller), or you can use a different typeface.

The easiest way to apply character changes is to use the Font Group of the Home tab of the Ribbon.

For this lesson we will use a document that you downloaded from the website.

Toward the beginning of the book there is a page that has Important Notice at the top. You will need to download the files from the website to access them for the lessons. If you have not downloaded the files go back to the Important Notice page and download the files.

Open the WordPad program if it is not open

Click the WordPad Menu Button and choose Open, or click the Open icon on the Quick Access Toolbar

The file we want to open is with all of the files that you downloaded. This is the first time we have tried to open a document that is in a different folder. Figure 4-30 shows the Open Dialog Box and the location of the folder where the files are located, provided that you put them in the My Documents folder.

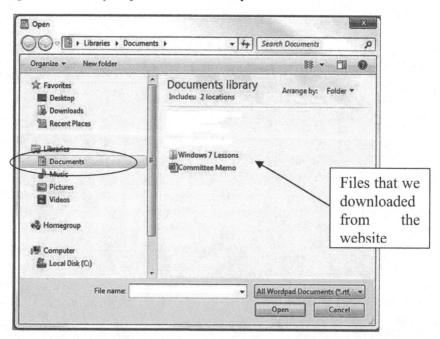

Figure 4-30

Double-click on the Windows 7 Lessons folder to open the folder

Click on Agenda **and then click the Open button**

The Agenda document will open and be on your screen. The Agenda screen is shown in Figure 4-31. We can use this document to explore some the different formatting options available.

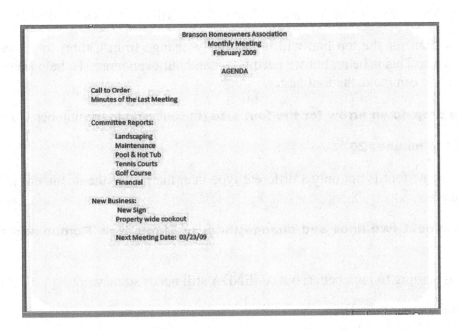

Figure 4-31

We will use the Font Group to make some simple formatting changes. This group is found on the Home tab of the Ribbon and is shown in Figure 4-32.

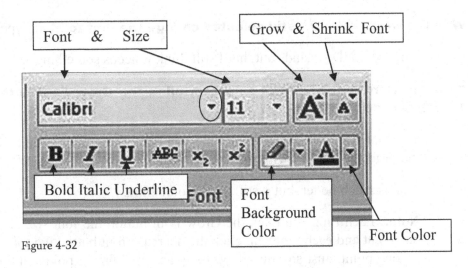

Figure 4-32

First we will change the typeface to something different than the default font.

Select the text in the first line of the document

Click the drop down arrow so we can change the font we are using (where the circle is in the figure)

Find Times New Roman **from the list and click on it**

The fonts are listed in alphabetical order, so it will down towards the bottom.

The font for the top line will immediately change from Calibri to Times New Roman. This is better but we need it to stand out even more. To help accomplish this we can make the font larger.

Click the drop down arrow for the font size (the one next to the number 11)

Click on the number 20

Now the font is not only a different type than the rest of the document it is also larger.

Select the next two lines and change them to Times New Roman and a size of 14

It is starting to look better, but AGENDA still needs some work.

Select the word AGENDA

Click the icon for Italic

Agenda will change to *Agenda*. That is not quite what I wanted, so click the Italic button a second time to put AGENDA back the way it was.

With the word AGENDA **still selected change the font to** Monotype Corsiva

Now AGENDA stands out, but I still think it needs something.

With the word AGENDA **still selected, click the Grow Font button until the Font Size is 16**

The font size will change each time you click the grow button and you can see the results in the box with the number in it.

That is much better, but it still needs something.

Note: Each time you click the Grow Font button the font size will increase by one point and each time you click the decrease font button the font will decrease by one point. Just so you know, fonts are measure in points and each point is 1/72 of an inch.

With the word AGENDA **still selected, click the underline button**

Now I am happy, it looks good. Say, what if we changed the color of the font to make it stand out even more?

With the word AGENDA **still selected, click the down arrow on the Font color button**

From here we can choose any of the theme colors or the standard colors. If the color you want is not shown you can click on the More Colors choice at the bottom.

Click on the Red choice from the Standard Colors

I thought that I was happy before but this is even better.

Note: Just so that you know, there are keyboard shortcuts for most of the items dealing with formatting on the Ribbon. Most of these use the CRTL key in combination with another key. Bold is CTRL + B, Italic is CTRL + I, Underline is CTRL + U.

We will want to save our changes in the document. What we **don't** want to do is save the changes in the original location which is the Windows 7 Lessons folder. You were asked to save all of your documents in the "MY Documents" folder.

To save the document we will need to click the WordPad Menu button and choose the "Save As" choice. Remember when I said that this allows us to save the document under a different name? I might need to clarify this a little. You cannot have two files in the same folder with the same name. You can have two files with the same name if they are in different folders or locations. We can save this in the My Documents folder and keep the same name.

Click the Menu button and choose Save As

The Save As dialog box will come to the screen and look similar to Figure 4-33.

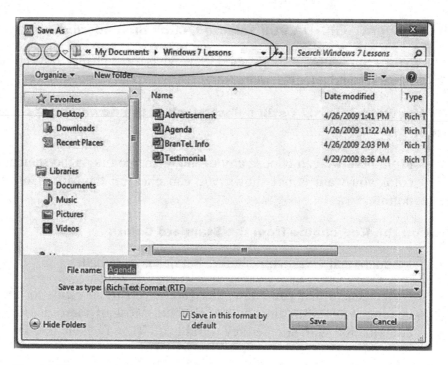

Figure 4-33

You can see from the figure that WordPad wants to save the file in the same place that it was when you opened it. We don't want this to be saved to the Windows 7 Lessons folder. We want it to be saved in the My Documents folder in the Document Library.

To make this happen, we will need to click on the small arrow to the left of Documents and then click on the My Documents folder. Figure 4-34 shows this.

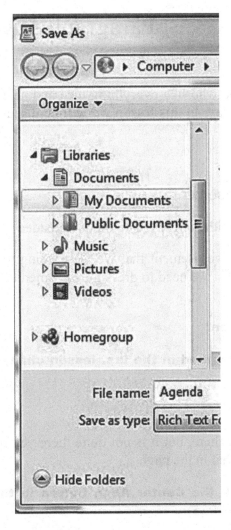

Figure 4-34

Once we have the My Documents folder selected, we can save the document. We can also leave the name the same since it is a different location from the original. To finish saving the document all we have to do is click the Save button.

Click the Save button

Lesson 4 – 12   Paragraph Alignment

If you are one of those people who like everything nice and neat and even, this lesson just might make your day. In this lesson we will be discussing how your text is aligned on the page.

If necessary open WordPad

Open the document titled Advertisement

This document can be found with the files that you downloaded.

This document contains an advertisement that we were going to run in a local magazine. We have decided that we need to dress the document up a little before we submit it to the magazine.

Select the first line of the document

Using the techniques you have learned in the last lesson change the font to Arial and the size to 18

Change the font color to Red

The first line is starting to get there, but it is not quite there yet. It would look a lot better if the text was centered in the page.

With the text still selected click the Center Align button in the Paragraph Group of the Home Tab (see Figure 4-35)

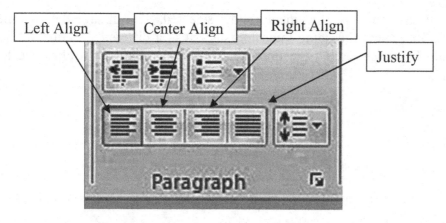

Figure 4-35

The text will now be centered on the page and will look much better.

Select the following words in the next line and make them bold

Quality

Dedicated

Hard working

In the next paragraph select excited and motivated **and make both words bold**

Select the last paragraph

The first paragraph in this document is center aligned on the page. The second paragraph is aligned to the left. Now we will experiment with the forth (last) paragraph.

Click on the Justify button

This will align the text on both the right and the left. Extra spaces will be added as necessary to accomplish this. This will keep both sides of the paragraph nice and even and it looks like you spent a lot of time making sure everything was perfect.

One of the things that drive me absolutely crazy is not having the text aligned on both sides.

As in the previous lesson, there are keyboard shortcuts that can be used. And again the majority of the shortcuts use the CRTL key in combination with other keys, some of these include CTRL + L for the left-align. CTRL + E is used for the center-align command. CTRL + R is for the right-align command. CTRL + J is for the justify command.

Save your changes (in My Documents) and then close the document

Lesson 4 – 13   Cut, Copy, and Paste

This lesson and the next lesson are dedicated to moving text and objects around in your Document. We will be using Cut, Copy, and Paste to accomplish this. First, let me give you a brief description of these commands.

Cut: This will <u>remove</u> any data that is highlighted (selected) and move it to the clipboard. The clipboard is a temporary storage place and will temporarily hold the data until you can put it someplace else in the document or even into a completely different document or program.

Copy: The copy command will <u>make a copy</u> of the selected text (or object) and place it in the clipboard for you to use later.

Paste: The paste command will copy from the clipboard and put the information into the document.

Once you have selected the text, you can move it to another place in the document by cutting and then pasting it elsewhere. Cutting and pasting text is one of the most common things you will do in WordPad.

The Clipboard is available from any program in Windows, so you can cut text from one program and paste it into another program.

If necessary open WordPad

Open the document titled BranTel Info

This document is located with the files that you downloaded from the website.

On the second page of the document we want to change the order of two of the paragraphs. We can use cut and paste for this.

Select the second paragraph on the second page, the one that deals with certification, and the blank line below the paragraph

When this is selected it will be highlighted in blue (See Figure 4-36). After you have selected the text it can be cut from the document.

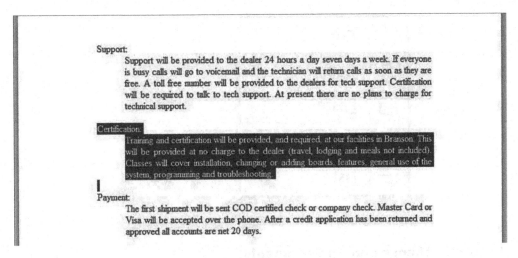

Support:

Support will be provided to the dealer 24 hours a day seven days a week. If everyone is busy calls will go to voicemail and the technician will return calls as soon as they are free. A toll free number will be provided to the dealers for tech support. Certification will be required to talk to tech support. At present there are no plans to charge for technical support.

Certification:

Training and certification will be provided, and required, at our facilities in Branson. This will be provided at no charge to the dealer (travel, lodging and meals not included). Classes will cover installation, changing or adding boards, features, general use of the system, programming and troubleshooting.

Payment:

The first shipment will be sent COD certified check or company check. Master Card or Visa will be accepted over the phone. After a credit application has been returned and approved all accounts are net 20 days.

Figure 4-36

Click on the Cut command (See Figure 4-37)

Figure 4-37

As you remember, the cut command is on the Home Tab of the Ribbon and in the Clipboard Group. The command is identified by a pair of scissors. As soon as you click the cut button, the text disappears from the document. The text is not really gone; it has been placed on the clipboard and is now ready for us to paste it to a new location.

Click the mouse at the beginning of the word Support

Click the Paste button

The text will magically jump back into the document where the insertion point was located.

Note: You can also use the right-click shortcut to cut and paste the text. If the text is highlighted and you right-click on it, the following shortcut menu will appear on the screen.

Figure 4-38

Save your changes (in My Documents)

There may come a time when you want some text to appear in more than one place in your document. Right now you probably can't think of a single reason why you would want this. You may not to reproduce an entire paragraph, but there might be a catch phrase that you want to use several times. In this part of the lesson we will actually reproduce an entire paragraph and have it appear in two different places in the document.

We are going to repeat the paragraph under Support for this part of the lesson.

Click your mouse at the very end of the paragraph about support and then press the Enter key

This will start a new paragraph.

Type the following and then press Enter

Let me repeat that for you:

We are setting up a place for the repeated text to appear. When we pressed Enter a new paragraph is started and this is where we will insert the repeated text.

Using the mouse select the entire paragraph that starts with "Support will be provided"

This is the text that we will copy.

Click the Copy button

You will notice that nothing seemed to happen when you clicked the copy button. Before, when we clicked the Cut button, the text disappeared from the screen. This time we only copied the text. That means that the original text is still in place and a copy of it was sent to the Clipboard.

Click the mouse on the line below the text you just typed

Click the Paste button

The text was moved from the Clipboard to the place where the insertion point was located. Remember the insertion point is the flashing vertical line. That is why we had to click the mouse just below the text we typed.

To end the lesson we will recap:

Cut: This will <u>remove</u> any data that is highlighted (selected) and move it to the clipboard. The clipboard is a temporary storage place and will temporarily hold the data until you can put it someplace else in the document or even into a completely different document or program.

Copy: The copy command will <u>make a copy</u> of the selected text (or object) and place it in the clipboard for you to use later.

Paste: The paste command will copy from the clipboard and put the information into the document.

Note: Don't forget there are keyboard shortcuts for these commands also. The shortcut for cut is CTRL + X, copy is CTRL + C, and paste is CTRL + V.

Save your work (in My Documents) and close the document

Lesson 4 – 14   Drag & Drop

In the last Lesson we learned how to cut text and paste it someplace else in our document. In this lesson we will learn an easier way to perform the same task. The method we are going to learn is called Drag and Drop. It is called this because we are going to select a section of text, drag it from its current location and drop it someplace else in the document.

Open the Committee Memo **document**

We will use this document to practice dragging and dropping text. I think you will enjoy moving text using this method. One reason is because it is so easy and the other reason is we all like having the power to command things to move.

Select the last paragraph

To drag the selected text we need to "grab" the text with the mouse.

If you move the mouse inside the highlighted area and then click and hold the left mouse button down, you will "grab" the selected text so that it can be moved. To move the text, we need to simply move the mouse and the selected text will follow.

There is something you need to watch for as you move the mouse. A small vertical line will move along with the mouse. This will show you where the text will be inserted when you release the left mouse button. In Figure 4-39 you can see that the text will be inserted just before the word Bill if I release the left mouse button.

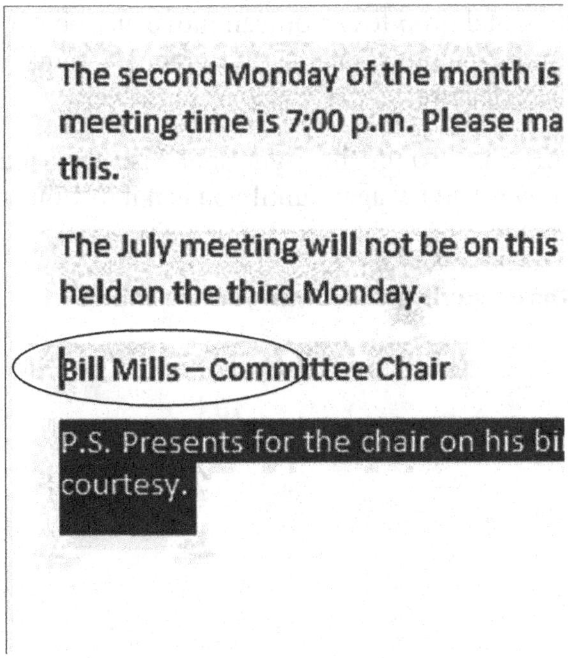

The second Monday of the month is
meeting time is 7:00 p.m. Please ma
this.

The July meeting will not be on this
held on the third Monday.

Bill Mills – Committee Chair

P.S. Presents for the chair on his bi
courtesy.

Figure 4-39

Using the mouse grab the selected text and drag it up and then drop it just before the word Bill

This may require a little practice but once you master this you can save a lot of time. If you did this correctly, your document should look like Figure 4-40.

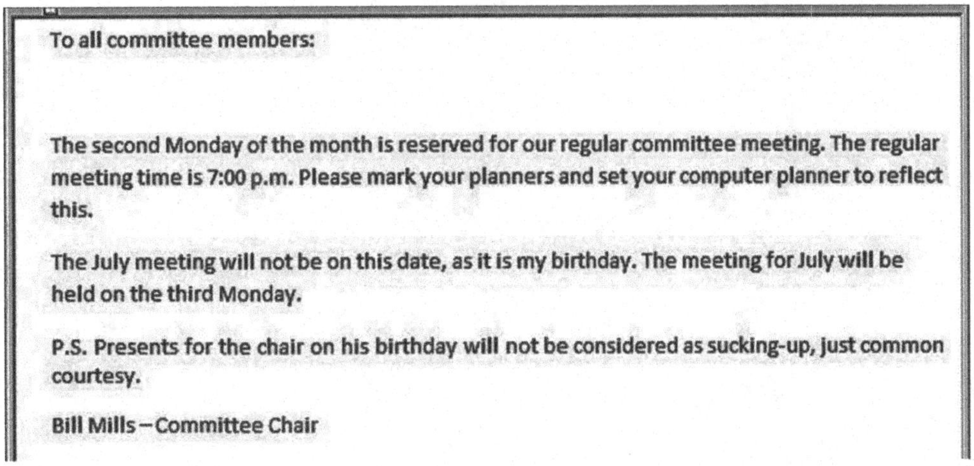

To all committee members:

The second Monday of the month is reserved for our regular committee meeting. The regular meeting time is 7:00 p.m. Please mark your planners and set your computer planner to reflect this.

The July meeting will not be on this date, as it is my birthday. The meeting for July will be held on the third Monday.

P.S. Presents for the chair on his birthday will not be considered as sucking-up, just common courtesy.

Bill Mills – Committee Chair

Figure 4-40

Not only can you drag and drop text, you can move any object in the document this way. You can move pictures, charts, and graphs. I think you will learn to love the Drag and Drop feature.

If had trouble getting the text in the right spot, use the Undo button to put everything back as it was and try again until you can move the text to the correct location.

Close the document without saving the changes we made

Lesson 4 – 15   Finding and Replacing Text

Searching for text in your document can be time consuming and frustrating. You know that somewhere in your document you referenced a certain word, but now you are trying to find it. Oh Boy! How much fun is this?

You could search through the entire document, word after word and page after page, or you could let WordPad do the search for you.

Open WordPad if it is not open

Open the Testimonial **document that is with the downloaded files**

Click the Find Button

The Find button is on the right end of the Home Tab, in the Editing Group.

The Find Dialog box will come onto the screen (See Figure 4-41). The Insertion point will already be positioned in the "Find what" text box ready for you to type in a word you want to find.

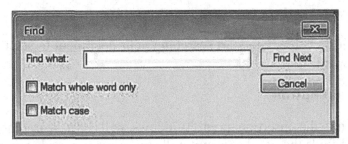

Figure 4-41

Using the keyboard type the word completely **and press the Enter key**

Windows will search the document and stop on the word completely. The testimonial would sound better if it said extremely pleased instead of completely pleased. By the way, the dialog box will stay on the screen until you close it just in case you want to search for another word.

Click the Cancel button in the dialog box

Replace the word completely with the word extremely by typing extremely

The word completely is already highlighted from the search so you do not have to do anything except start typing to replace the word.

You can also have Windows automatically replace words for you.

Click on the Replace command

This command is just below the Find command in the Editing Group.

The Replace dialog box comes onto the screen (See Figure 4-42).

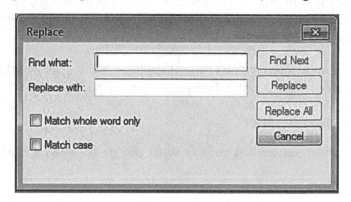

Figure 4-42

In this example we want to replace the word screen with the word monitor.

In the Find what textbox type the word screen

In the Replace with textbox type the word monitor

Now Windows is ready to start searching your document to see if it can find the requested word.

Click the Find Next button

Starting from the insertion point Windows will start searching for the word screen. If it finds the word screen it will stop searching and stop on the word. At this point you can decide if this is the instance of the word screen that you want to replace. If it is, all you have to do is click on the Replace button.

Click the Replace button

The word screen has now been replaced with the word monitor. Windows 7 will now start looking for the next occurrence of the word screen or you can jump ahead and choose to replace all occurrences of the word.

A word of caution: If you click on the "Replace All" button, Windows will replace every occurrence of the word screen with the word monitor. Be very careful this may not really be what you wanted to accomplish.

Click the Cancel button to remove the dialog box from the screen

Save the document (in My Documents) and then close it

Lesson 4 – 16   Printing a File

Let's recap: We can enter text, save the file, close the file, and open a file. Now let's see about printing a document. The document may be part of a report you are submitting or it may just be your personal letter. Either way you will probably want a printed copy. Printing the document is a simple process, provided that you have a printer set up on your computer.

Open WordPad if it is not open

Open the document named Committee Memo

The document should be displayed on the screen and look like it did when we last worked on it.

Click the WordPad Menu Button and move the mouse to the Print choice, but do not click the mouse

There are three choices you can make from this menu: Print, Quick Print, and Print Preview. Before we actually print the document, it might be a good idea to see how it will look when we print it. We may find out that it will not all fit on one page and we may have to do some adjusting to our sheet. The available print choices can be seen In Figure 4-43.

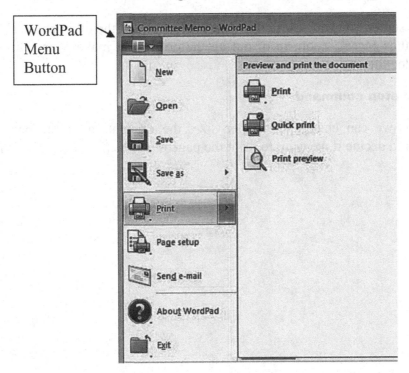

Figure 4-43

Click on the Print Preview choice

This will display how the sheet will look when it is printed. There are also a few other options available from this screen (see Figure 4-44).

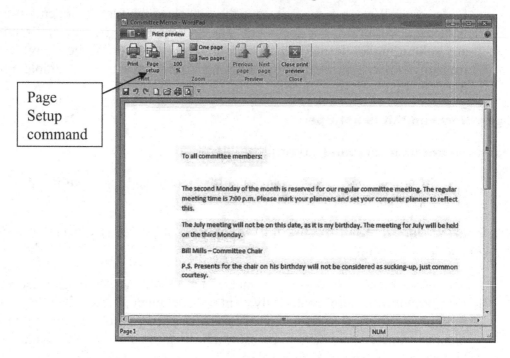

Figure 4-44

From this screen you can: Close the print preview, view the other pages (if there are any other pages), Zoom in or out on the document, make page settings, or print the document.

Click the Page Setup command

From here we can change the paper size, the orientation of the paper, the margins, and decide if we want to print the page numbers (See Figure 4-45).

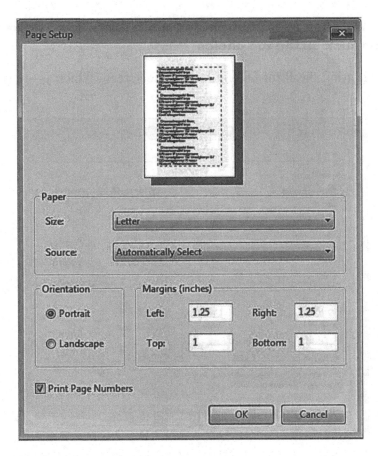

Figure 4-45

Letter is going to be the most common size of paper used, however you could also choose from the other 44 sizes that are available.

The Orientation decides how the text is printed. The page could be Portrait with the long side of the paper on the right and left or it could be Landscape, with the long side of the paper being across the top and bottom.

Click the radio button beside Landscape and watch the preview area at the top

Click back on Portrait and note the difference

You could also choose to change the margins of the page. You could have the margins closer to the edge of the paper or further away.

Click the Cancel button and then click the Close print preview button at the top

Now that we have seen what it is going to look like when we print it, let's see what the other two choices are when we moved the mouse over the print option.

Click on the WordPad Menu button and move the mouse down to the word Print and then click on the top choice: Print

This will bring the Print Dialog box to the screen, which is shown in Figure 4-46.

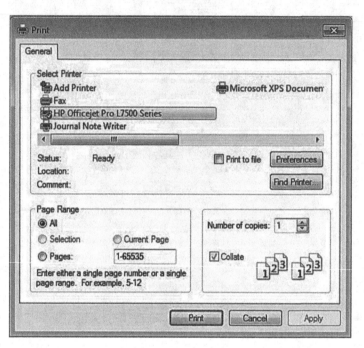

Figure 4-46

The Print Dialog box will allow us to choose the printer we wish to use by clicking on the name of the printer we want to use under the Select Printer section assuming, of course, that there is more than one printer configured on your computer.

You can also choose to print all of the pages in the document, or the current page, or you can specify which pages to print. You can also chose to collate the pages (put them in the correct order as they print). If we click OK the document will print. If we click Cancel the Print Dialog box will go away and the document will not print.

In addition to the above mentioned choices we can also choose to print only the selection that we have highlighted. This choice will become available if we actually have some text highlighted. On the right side we can choose how many copies we want to print.

Once we have everything marked correctly, all we have to do is click on the OK button and this will send the information to the printer.

Click Cancel so nothing will print

The last choice, of the three choices that came up when we moved the mouse to the Print command after we clicked on the WordPad Menu Button, is the Quick Print choice and if you click on this choice the entire document will be sent to the default printer. There will be no dialog box for you to make choices. The document will just be sent to the printer.

Note: If you click on the print choice from the WordPad Menu Button, the Print Dialog box will come onto the screen. This is identical to the top choice of the three available choices we just discussed.

Close the WordPad program

Lesson 4 – 17　Keyboard Shortcuts

As we have gone through this chapter, we have continued to mention the keyboard shortcuts available to you. In this chapter I will present the most commonly used keyboard shortcuts.

Below are tables showing some of the available keyboard shortcuts. A complete list of all of the shortcuts is provided for you on your computer in the Help section. To find the list click on the Help button (the small question mark in the upper right corner of the screen) and type keyboard shortcuts in the search window. Some of that list has been reproduced for you in the table below.

Table 4-1 - Some of the available Shortcuts

To do this	Press this
Delete one character to the left of the insertion point	BACKSPACE key
Delete one word to the left of the insertion point	CTRL+BACKSPACE (actually press Ctrl and then press Backspace and then release both keys)
Delete one character to the right of the insertion point	DELETE key
Delete one word to the right of the insertion point	CTRL+DELETE (actually press Ctrl and then press Delete and then release both keys)
Copy selected text or graphics to the Office Clipboard	CTRL+C (actually press Ctrl and then press C and then release both keys)
Cut selected text or graphics to the Office Clipboard.	CTRL+X (actually press Ctrl and then press X and then release both keys)

Paste the most recent addition to the Office Clipboard.	CTRL+V(actually press Ctrl and then press V and then release both keys)
Make letters bold.	CTRL+B (actually press Ctrl and then press C and then release both keys
Make letters italic.	CTRL+I (actually press Ctrl and then press C and then release both keys)
Make letters underline.	CTRL+U (actually press Ctrl and then press C and then release both keys)
Undo the last action.	CTRL+Z (actually press Ctrl and then press Z and then release both keys)
Redo the last action.	CTRL+Y (actually press Ctrl and then press Y and then release both keys)
Switch a paragraph between justified and left-aligned.	CTRL+J (actually press Ctrl and then press J and then release both keys)
Switch a paragraph between right-aligned and left-aligned.	CTRL+R (actually press Ctrl and then press R and then release both keys)
Switch a paragraph between centered and left-aligned.	CTRL+E (actually press Ctrl and then press E and then release both keys)
Left align a paragraph.	CTRL+L (actually press Ctrl and then press L and then release both keys)

Remove paragraph formatting.	**CTRL+Q** (actually press Ctrl and then press Q and then release both keys)
Get Help or visit Microsoft Office Online.	F1 Key
Show Key Tips.	Alt key

Table 4-1

As you can see there are several shortcut keys available for you to use. This is not even a complete list of all of the shortcut keys that are available. If you still need more commands, click on the help button and search for Keyboard Shortcuts. The Help button is the small question box at the top right side of the screen.

Chapter Five   Managing Files

This chapter is not really a part of Microsoft Windows 7. It is included in this book because it is so important. After you create your documents you will need to save them in a place that will enable you to find them easily when you need them.

If you are one of those people who have your entire monitor screen filled with icons and your documents folder is so full you have trouble finding the correct file, this chapter may change your computer life.

In this chapter we will learn how to make folders and how to move your documents around so that each document gets in the correct folder. You will learn how to organize your computer.

Just so that you remember, these documents we have been making are actually files that you are storing on your hard drive.

Lesson 5 – 1　　　Making New Folders

If you were at home storing papers you probably wouldn't put them in a big pile in the center of your table or desk. You would more than likely put them in a filing cabinet. In your filing cabinet you would not just randomly push papers into any drawer. You would have several folders to keep similar documents together, and have the papers all neatly stacked inside each folder. This is the same line of thought you need to have with your computer.

Before you can organize your files, you need a place to keep them. Our first lesson is dedicated to teaching you how to make folders.

Click on the Start button and select Documents on the right side

The Windows Explorer program will jump onto the screen. You will notice that you are in the Documents Library and you will probably have two folders and some documents showing on the screen. Figure 5-1 shows an example of what your screen might look like.

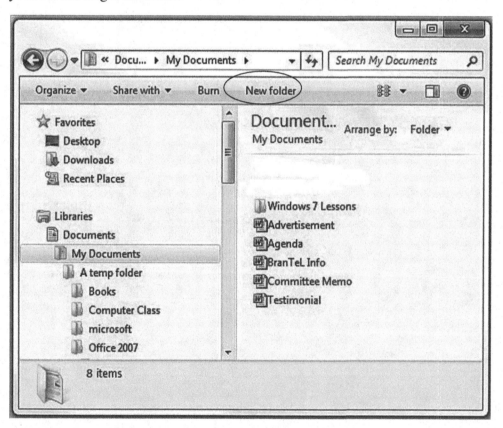

Figure 5-1

There are, as usual, a couple of different ways to add a new folder to the Documents Library. We will look at the most common way first.

Click on New Folder by the top

Figure 5-2 shows the results of clicking on the New Folder button.

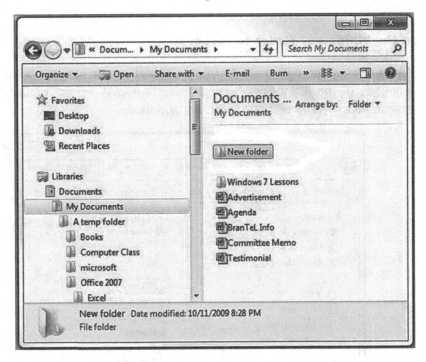

Figure 5-2

The name "New Folder" has been highlighted automatically by Windows because we will want to give the folder a different and unique name.

The original name of the folder will now be highlighted and ready for you to start typing a different name in the textbox. You do not have to click the mouse or anything; just start typing a new name.

Using the keyboard type the following name and press Enter when you are finished

Windows 7

The folder will now have the new name on it. Now you can start to organize all of your files.

You may want a folder to hold all of your letters to mom or dad. You may want to have a folder to hold that famous Christmas letter that you write every year. You may want a folder to hold your recipes. Think of the folders you might need to organize your documents.

Go ahead and make some folders that you might need.

> **Note:** By the way, the other way to create a new folder was to right-click inside the Document area and select New and then Folder from the drop down menu.

> Now that you have at least one new folder, we need to see what is inside the folder.

Click on the Windows7 folder to select it

> There are a couple of ways to open a folder so we can see what is inside of it. There is no hard way, so here are the two easy ways. Figure 5-3 shows the screen you will see on your computer monitor.

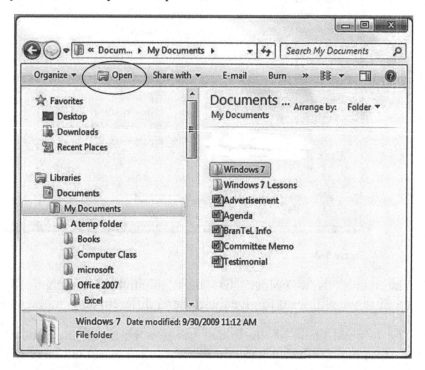

Figure 5-3

> The first way is to click on the Open button (the one circled in the figure). The other way is to double-click on the folder itself. Either way you will get the same results. You will also want to notice that the Open button on the toolbar has replaced the New Folder button. When you click on a folder, you cannot add another folder until you either open the folder you are selecting or unselect the folder.

Click the Open Button

> Your screen will change to show you what is inside of the Windows7 Class folder. As you can see from Figure 5-4, the folder has nothing in it.

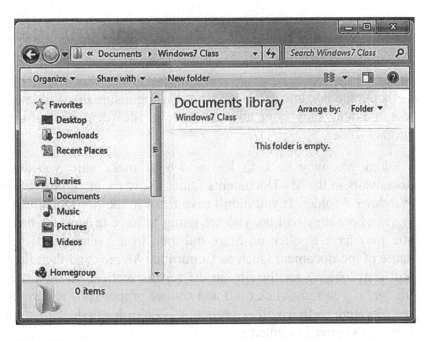

Figure 5-4

Well that doesn't do us a lot of good. I can remember the first computer I bought. It had those big 5 ¼ inch big floppy disks. (That's not so bad; the first one I ever used had cassette tapes in it). Anyway, I followed all of the instruction on adding folders to the disk and when I got all the folders on the disk; that was as far as it went. I called the manufacturer and asked them what to do next because I couldn't see any advantage to having empty folders. The person on the phone told me they didn't know what to do next or how to put something inside the folders. In the next lesson we will solve that problem.

Click the Close button in the top right corner of the window

Lesson 5 – 2   Moving Files

Let's put the documents we have been using inside the Windows7 folder. In this lesson I will show you how to move a file (document) from one folder to another folder.

I asked you way back in lesson 4-6 to make sure you saved all of your documents in the My Documents folder. Now we are going to move them to the Windows 7 folder. If you didn't save them in the My Documents folder, I don't know where they will be, you are going to have to look for them. To find them you may have to click on Start and then in the search area you can enter the name of the document (such as Committee Memo) and then the search program will start looking for the file. In the results section, you will be able to see the file and if you right-click on it and choose properties you can see where the file is located. It will probably be located in something like C:\Users\Owner\Documents

The rest of this lesson is going to assume that your documents are located in the My Documents folder.

Click on the Start button and then click on Documents.

You should see, somewhere in there, a set of files that look something like the ones shown in Figure 5-5.

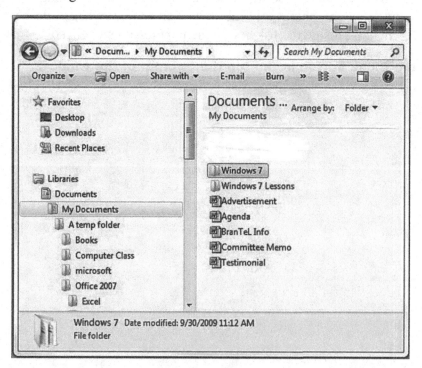

Figure 5-5

156

You should have on your computer the 5 documents that we have used in the lessons and a folder that has Windows7 on it.

We want to move these 5 documents to the Windows7 folder. Let's move the Agenda document first.

Right-click on the "Agenda" document and choose Cut from the drop down list

Double-click on the Windows7 folder to open it

The screen should change and show the contents of the folder. There are no contents and the folder should be empty.

Right-click inside the blank part of the right side of the screen and click on Paste from the drop down list

Your screen should now look like Figure 5-6.

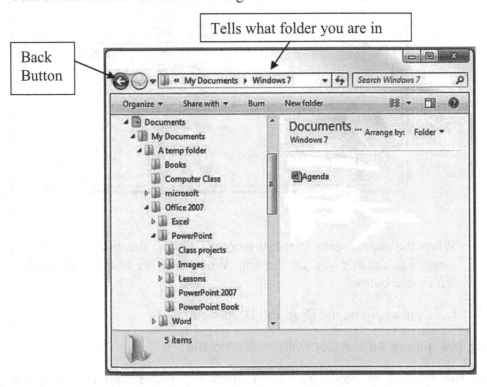

Figure 5-6

The folder is no longer empty, it has one file in it; the Agenda document.

Click the Back button and see that the Agenda file has been moved

As I normally tell you there are easier ways to do these things; that was the hard way.

You still have four documents (files) that you need to move. Do you remember the Drag and Drop method we used earlier? I hope so, because we are going to use it to move the other files.

Click on the BranTel Info **document (file) and drag it to the Windows7 folder**

Figure 5-7 shows the screen as you move the file to the folder.

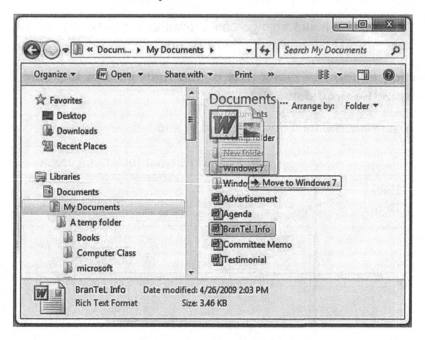

Figure 5-7

When the mouse gets to the Windows7 folder, the folder will be highlighted along with the file you are moving. When they are both highlighted release the left mouse button.

Let's put a spin on the Drag and Drop method.

Click the mouse on the Committee Memo **file**

Hold the CRTL key down and then click the mouse on the Testimonial file and then click the mouse on the Advertisement file now release the CTRL key

All three files should be highlighted. Holding the CTRL key down will allow you to select more than one object, in this case more than one file. You can now move all three files to the Windows7 folder at one time with the Drag and Drop method.

Click the mouse anywhere inside of the highlighted area and drag all of the files to the Windows7 folder. When the Windows7 folder is highlighted release the left mouse button

Double-click on the Windows7 folder to open it and view the contents

All 5 files should now be in the Windows7 folder.

Click the Close button on the window

Lesson 5 – 3   Copying Files

Just for the sake of having something to do, we want to keep our documents in the Windows7 folder, but we also want the "Committee Memo" file to be in the My Documents folder. To accomplish this we will need to copy the file from one folder to another, not move it.

Using the method you have learned before open the Windows7 folder

Your screen should look similar to Figure 5-8.

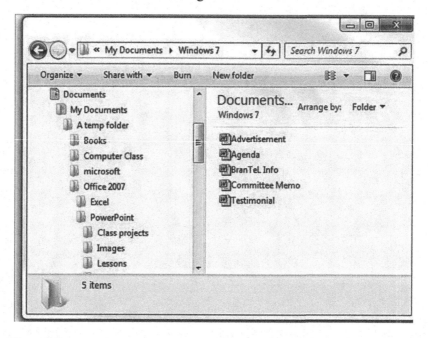

Figure 5-8

We want to copy the Committee Memo document to the My Documents folder. We also want the original to remain in the Windows7 folder.

Click on the Committee Memo file

Again, there is more than one way to do this. You could right-click on the file and then choose Copy from the drop down list. Let's try something we have not tried yet.

Press the ALT key and then release it

This will temporarily bring the old Menu bar back to the screen. Figure 5-9 shows the menu bar at the top of the Windows Explorer window.

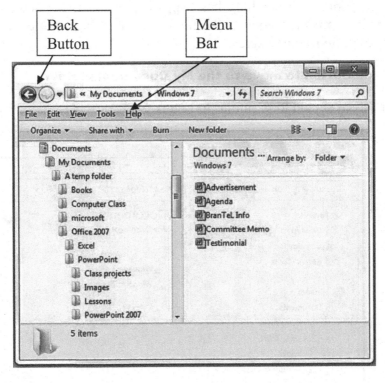

Figure 5-9

Click on Edit from the Menu Bar and then click on Copy

Figure 5-10 shows the drop down menu from the Edit button on the Menu bar.

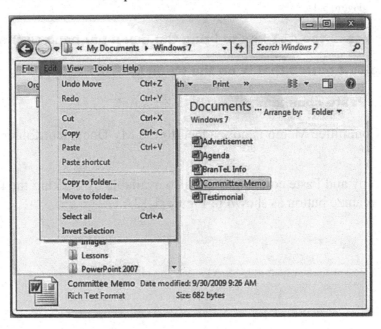

Figure 5-10

It doesn't look like anything has happened as far as the screen is concerned. What did happen is a copy of the file has been placed on the Clipboard waiting for us to Paste it somewhere.

Click the Back Button to move to the My Documents folder

Your screen should look similar to Figure 5-11.

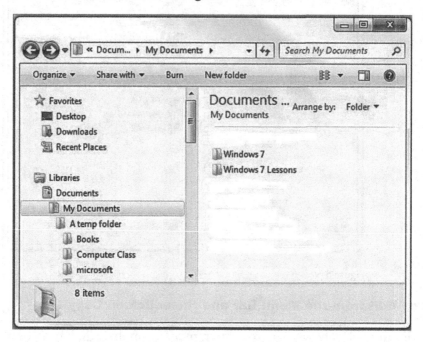

Figure 5-11

Press and release the ALT key to bring the menu bar back to the screen

Click on the Edit button that is on the Menu bar

Click on the Paste command

The Committee Memo is now added to the My Document folder (See Figure 5-12).

The copy and Paste commands are also available by clicking the Down arrow on the Organize button as shown in Figure 5-12A.

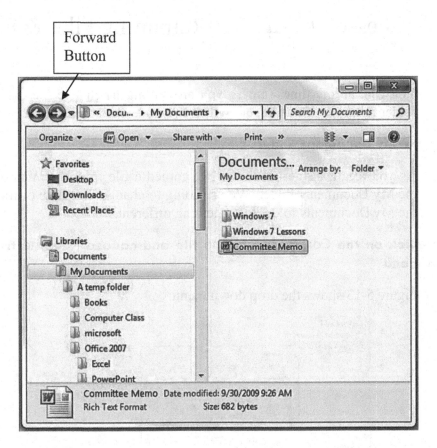

Forward
Button

Figure 5-12

If you want, you can click on the Forward Button to make sure the file is in both locations.

Figure 5-12A

Close the window by clicking on the Close button

Lesson 5 – 4   Renaming Files & Folders

You can, at any time, (unless you are editing it) change the name of a file or folder. This very short lesson will show you how to do just that.

Open the My Document folder

As you recall, in the last lesson we copied a file from the Windows7 folder to the My Documents folder. We are going to change the name of the document in the My Documents folder to something different.

Right-click on the Committee Memo **file and choose** Rename **from the drop down menu**

Figure 5-13 shows the drop down menu.

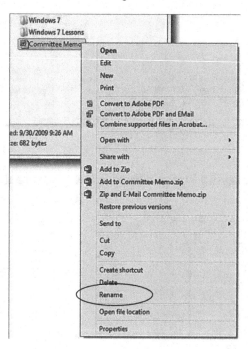

Figure 5-13

The original name (Committee Memo) will be highlighted and ready for you to type in a new name for the file.

Using the keyboard type Short Memo **and then press Enter**

The name is now changed. The new file name is now Short Memo.

This technique can be used to rename a file or a folder.

Close the My Documents window

Chapter Six   The Internet

The internet is a large part of our society today. Normally I would go into all of the boring tales of how the internet was started and how it has evolved into what it is today. But I imagine that you are more interested in how to use the internet than where it came from. This chapter is dedicated to showing you how to use the internet.

You will see, as we go through this chapter, that some of the lessons will intertwine. I am not usually a big fan of doing that, but it will be easier on you to do this chapter that way.

Lesson 6 – 1   Internet Explorer

There are three things you need to be able to get to the internet (besides your computer).

First you need an Internet Service Provider (ISP). This is the company that lets you use their software to connect to the internet. These include such companies as AOL, Net Zero, the local cable company, or even the local telephone company.

Second you need a modem, this could be a cable modem, a DSL modem, or even the older dial-up modems. This allows your computer to "call in" to the ISP.

The third thing that you need is a web browser. There are several web browsers available and include Internet Explorer (which comes standard on your computer). There are also third party browsers such as Safari and Firefox, which are both free, on the internet.

Open Internet Explorer

Internet Explorer might be located on the Taskbar or you might see it when you click the Start Button. If you still do not see the program, it will be towards the top of the programs under All Programs.

The first time you open Internet Explorer you are going to get a Welcome Screen asking you to turn on some of the new features. Figure 6-1 shows the Welcome Screen.

Figure 6-1

It will be okay to click the Next button.

Click the Next button

The next screen will ask you to turn on the feature that will allow Windows to suggest web sites that you might like based on the previous websites you have visited.

If you don't want this feature turned on click the radio button next to No

If you want this, click the radio button next to yes

I turned mine on and so far my computer has not caught fire or blown up or anything.

Click the Next button

The next screen wants to know if Windows can make some of the choices for you. Personally I don't like anyone else making my choices for me, so I clicked the Choose custom settings.

Click the button next to Choose custom settings and then click next

This part of the setup will let you choose the default search provider. This does not mean that you cannot go to Google or Ask or any search provider that you want. This means that whenever you type something in the search area at the top (that says Live Search in it), do you want to use "Live Search" as the search provider? It will be okay to keep the current search provider.

Click on the small circle next to Keep my current default search provider and then click the Next button

The next screen wants to know if you want to get updates as they are available. You will probably want this.

Click the Yes choice and then click the Next button

The next page deals with Accelerators. We will discuss these more a little later. For now go ahead and click the Keep my current Accelerators.

Click on Keep my current Accelerators and then click the Next button

That brings us to the Compatibility Settings. We will discuss this a little later also.

Click the No choice and then click Finish

By the way, if you decide that you don't like these settings they can be changed back. We will discuss that later.

Now you will have Internet Explorer up and ready to go.

Internet Explorer 8 is the latest version of Internet Explorer. Internet Explorer looks a little different than some of the earlier versions and Figure 6-2 shows the top part of the window.

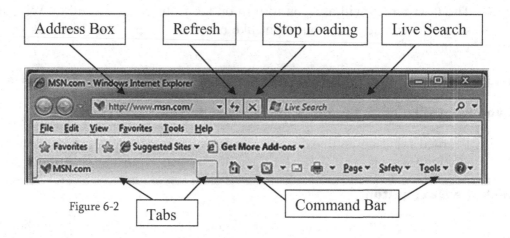

Figure 6-2

Across the top of the window is the Title bar. This shows the webpage that you are on and the name of the program you are using. On the far right side of the Title bar are the Minimize, Maximize, and Close buttons.

Directly below the Title bar is the Address Box. This is where you type the address of the web site that you want to visit.

At the end of the Address Box is the Refresh button. Clicking this will let you reload the current web page and get a fresh view of it.

This is followed by the Stop button. Clicking this will cause the current web page to stop loading. This is for when the web page takes forever to load and you just want to quit and go someplace else.

The last thing on this row is the Live Search area. This search feature will let you search for web pages from whatever web page you happen to be on. We will discuss this a little later.

The next row down may or may not be visible on your screen. This is the menu bar and is not shown by default. It can, however be turned on and we will discuss this in another lesson.

Below the Menu bar is the Favorites Bar which hosts two new features: the Suggested Sites and Get More Add-ons. These will also be discussed in later lessons.

Directly below the Favorites Bar, and on the left side, are the different Tabs. If I remember correctly, this feature came out in Internet Explorer 7. You can have several tabs open at the same time and each one can be on a different web site. We will discuss this more in the next lesson.

To the right of the Tabs is the Command bar. You would probably recognize the name better if they had left it as the toolbar. This is the home of the shortcuts to features that you will use over and over. We will discuss these later also.

This, in a nutshell, is Internet Explorer 8. We will spend the rest of the chapter using Internet Explorer 8 and explaining the new parts.

Lesson 6 – 2   Getting to a Web Page

Now that you have started Internet Explorer you are probably wondering "now what"? You obviously want to go somewhere on the internet, so how do you get there? If you know the web address, you can type it in the web address bar.

Click your mouse in the address box and type the following

www.google.com

Press the Enter key when you are finished typing

You will be taken to the home of Google. Google is a search engine that many people use to search for and find web sites.

It gets even easier. All you have to type is the last part of the web address. Microsoft will fill in the "http://www." for you.

Click the mouse in the address box again

The entire address should be highlighted. If it is not highlighted, click the mouse at the right end of the text and holding the left mouse button down drag the mouse to the left end of the text.

Type the following and press the Enter key when you are finished

bransonumc.org

You will be taken to the Branson Missouri United Methodist Church web site. It does get even easier, but we will cover that a little later.

There is always the possibility that you will find two web sites that you need to look at, and you are constantly going back and forth between the two. First you load one site and then you have to go back and load the other site, and then back again. Surely there is an easier way!

Earlier I mentioned that there were Tabs below the Favorites bar. Using the Tabs will solve this problem. We will take a few moments and look at the Tabs. Figure 6-3 points out the Tabs.

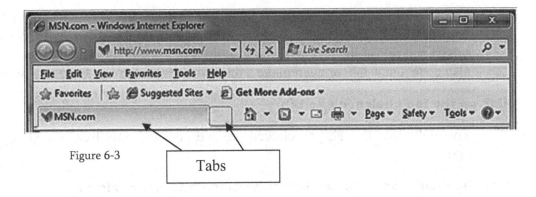

Figure 6-3

Tabs

Type google.com in the address box and then press the Enter key

Click the mouse on the second smaller tab

Type bransonumc.org in the address box and press Enter

Both tabs now a name on them indicating the web page they are connected to. There is also a third smaller tab visible. You could click on that tab and go to a third web site. If you did a forth tab would become visible.

Using the mouse, click on the Google tab and watch your screen change to show that web site

Now click on the Branson United Methodist tab and you will immediately go to that web site

You have two different web sites open and they do not interfere with each other. If you decide to close one of the tabs, you can close it without disturbing the other tab. Let's close the church web site.

Right-click on the Branson United Methodist tab

The shortcut menu shown in Figure 6-4 will jump onto the screen.

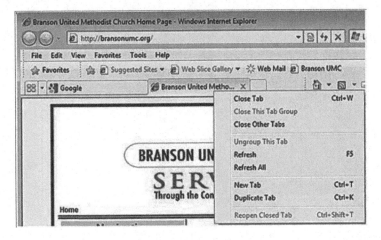

Figure 6-4

From here there are several things that you can do. You can close the tab that you clicked on, you could also close all of the other tabs, you could refresh this screen and reload the web page, you could also refresh all of the tabs, you could also go to a new tab, or even make a duplicate of this tab and have two of them.

Click on the top choice, close tab

This will close the tab you clicked on, and you will only have the Google tab open.

Click the mouse on the close button to close Internet Explorer

Closing Internet Explorer will allow us to try something out and see if it works. Do you remember way back in Lesson 1-7 when we added the Address Bar to the Task Bar? At that time I told you we would talk a little more about it later and now it is later.

Click the mouse inside the Address bar that we added to the Task bar

Type Google.com and then press the Enter key

Internet Explorer will open and the home page for Google will be displayed on the screen.

Click the mouse inside the Address bar that is on the Task Bar again

Type bransonumc.org and then press the Enter key

The second tab will open and the Branson United Methodist Church home page will be displayed on the screen. This is so much easier than opening Internet Explorer and then having to type the web address in the Address bar at the top. Windows does most of the work for us now. I think this will save you a great deal of time.

Close Internet Explorer

Lesson 6 – 3   The Home Page

How is it that whenever you start Internet Explorer it goes to a certain web site? This may not be the web site that you would like it to go to. In this lesson you will learn how to have your computer go to the website of your choice when you start Internet Explorer.

Open Internet Explorer if it is not open

When Internet Explorer opens it automatically goes to a certain web page. This is called the Home Page and every web browser goes to a Home Page. The Home page can be changed if you do not like the current Home Page. We will change the Home Page in this lesson.

The first thing we want to do is go to a different web page away from the Home Page.

Type ez2understandcomputerbooks.com **into the address box and press Enter**

Now we will make the Home Page change.

Click on the Tools button on the Command Bar

The Command Bar is shown in Figure 6-5.

Figure 6-5

A drop down menu will appear on the screen and is shown in Figure 6-6.

Figure 6-6

At the bottom of the menu is a command that says Internet Options, click on it

The Internet Options Dialog box will come onto the screen as shown in Figure 6-7.

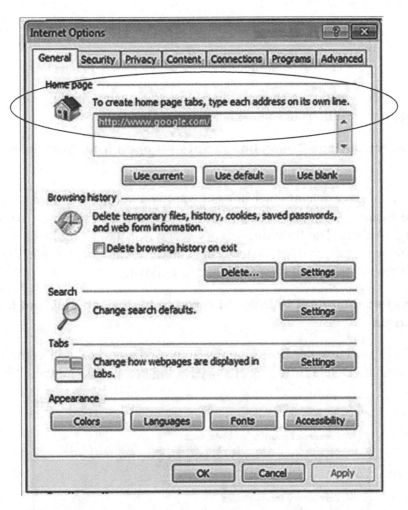

Figure 6-7

The top part of the dialog box is where we will change the Home page. If you know the full address of the page you want to use for the Home Page, you can type it in the space provided. If you happen to be on the web page that you want to use, you can click the "Use Current" button.

Click the Use Current button

The home page will be changed to
http://www.ez2understandcomputerbooks.com

Click the Apply button

Now every time you open the Internet Explorer program, you will be taken to the ez2understandcomputerbooks web site. If this is not the web page that you would want for your Home Page, go to the page that you want and repeat the above process to change your Home Page to the appropriate web page.

The next lesson will continue with the Internet options.

Lesson 6 – 4   More Internet Options

In the last lesson we changed the Home Page using the Internet options. In this lesson we will look at some other Internet Options.

If the Internet Options Dialog box is not still open from the last lesson open it now

There are a couple of other things I want you to look at before we move on from the Internet Options Dialog box. The first is the section just below the Home Page section. The browsing history section of the dialog box will allow you to remove temporary files, history, and cookies. Allowing too much stuff to pile up in these sections will slow your computer down.

Click the checkbox next to Delete browsing history on exit and then click the Delete button

When you click the Delete button another dialog box will come to the screen. This is the Delete Browsing History Dialog box and is shown in Figure 6-8.

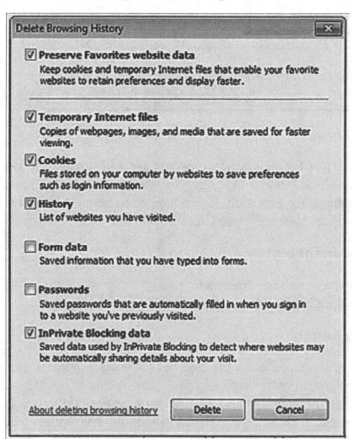

Figure 6-8

There are some things that you will want to delete and some other things that you will want to keep. When you look at the list, you might be tempted to say something like "these don't look so bad, I should probably keep these things".

The reality is that you should, every so often, delete the temporary internet files, the cookies, the history, and the InPrivate Blocking Data. The Form data and the saved passwords you might want to keep. These are no match for good record keeping, and I am sure that you are all very good at keeping a record of all your passwords, but these can save you some time and frustration. The top choice is one that could go either way. If you saved the web site as one of your favorites so you can go back to it often, you may want to keep those cookies.

Click each of the checkboxes that are clicked in the figure and then click the Delete button

You may see a small window on the screen for a few seconds while Windows deletes the files from your computer and then you will be back at the Internet Options dialog box.

The next thing I want you to look at is on the Privacy Tab.

Click the Privacy tab at the top of the dialog box

If it is not already there, move the slider to medium high

The default setting is set to medium. The Medium High is a better setting; at least I think it is. The medium setting will restrict some of the cookies and the medium high will block the cookies.

Make sure the checkbox next to Turn on pop-up blocker is checked

Pop-ups are those annoying advertisements that pop up on your screen when you go to some web sites. I try to block these whenever possible.

Click the OK button to save any changes that you made

This will also close the dialog box

Lesson 6 – 5   The Tool Bars

In lesson 6-1 I told you briefly about the Menu bar and that it might not be visible on your screen. In this lesson you will learn how to show and hide the different toolbars. There are several ways to do this and I am only going to show you a couple of the ways.

Click the Tools button on the command bar

This will bring the same drop down menu to the screen that we saw in the last lesson. This time we are going to be looking at the toolbar command.

Move your mouse down the list until you come to Toolbars and let the mouse hover there for a moment

Another menu will slide out to the side with a list of the available toolbars. If there is a checkmark next to the toolbar it is visible (See Figure 6-9).

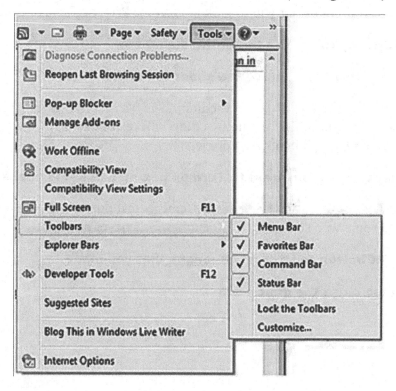

Figure 6-9

Try turning these off and on one at a time and see how they affect your screen

I think you will want to have all four of these checked when you are finished. By the way, each time you make a change, the menus will go away and you will need to start again by clicking the Tools button.

As long as we are talking about the different toolbars, we might as well look at some of the commands on the toolbars.

The Menu bar, shown in Figure 6-10, has just about every command available that you will need to control Internet Explorer. If you have ever used Windows before, you should be familiar with menu bars. Most of the commands on the Menu Bar will be covered as we look at the Command Bar.

Click on the File command and notice the choices that are available to you

Slowly move the mouse to the right and see the other commands

Figure 6-10

The Command bar, shown in Figure 6-11, has several shortcuts on it. These are shortcuts that you can use to reach some of the most popular commands that are found in the menu bar.

Figure 6-11

The small house, on the far left, is the Home command. Clicking on this will take you directly to your Home page.

Click on the Home icon

If you click on the down arrow next to it you can choose to add the current tab as a home page or change your homepage. This is going to require some explanation.

With Internet Explorer 7 some of the old rules have changed. In the olden days (a little over two years ago) you were allowed to have one Home Page. Now you are allowed to have several home pages. Not only that but each Home Page will open in a different tab when you start Internet Explorer. In lesson 6-3 I told you to set your Home Page to ez2understandcomputerbooks.com. Then we opened a second tab to the Branson Missouri United Methodist Church web site. Now we will use the second tab to create a second Home Page.

With Internet Explorer open and your Home Page displayed on tab 1, click on the second tab

Type bransonumc.org **in the address bar, if it not still there, and press Enter**

Once again you will have two tabs with a different web page showing on each one of them. The second tab should be on the front and have the information about the church on it.

Click the down arrow next to the small house (the Home Button/icon)

Click on Add or change home page

The Add or Change Home Page Dialog Box will come to the screen (See Figure 6-12).

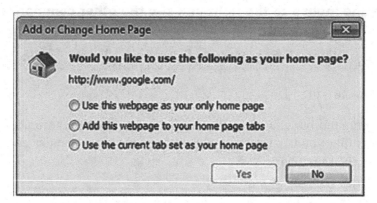

Figure 6-12

We can change our Home Page if we click on the "Use this webpage as your only home page" radio button.

We can add a second Home Page and have it show on tab two if we click the radio button next to "Add this webpage to your home page tabs".

We can change tab 2 to be the tab that shows the Home Page if we choose the last choice.

Click the small circle (radio button) of the center choice and then click the Yes button

Close Internet Explorer by clicking the mouse on the Close button (the red button with the X on it) and choose close all tabs

Open Internet Explorer

The first tab should have ez2understandcomputerbooks showing on it, or the page you changes it to if you did not like the choice we made earlier, and the second tab will have the church website on it.

Click on the second tab and make sure the second home page is there

If you do not want the second home page, I will show you how to remove it.

Click the down arrow on the Home button and move the mouse down to Remove

Click on the second choice (The Branson United Methodist Church) and the second home page will be gone as soon as you click the Yes button to confirm that you want to delete the second Home page

Having two Home Pages can be very convenient if you have more than one web page that you visit a lot.

In the off-chance that you had three Home pages when you started Internet Explorer and you don't want the Microsoft web site to come up as a Home Page (which it might very well have), I will show you how to get rid of it. If you did not have three Home pages on your screen, you still might want to see this so follow along.

Click the down arrow next to Tools and then choose Internet Options

This is the same dialog box we used to set the Home page to ez2understandcomputerbooks. At the top all of the Home pages are listed. If there is an extra Home page that has Microsoft on it, highlight it with the mouse and then click the Delete key on the keyboard. This will remove the extra Home page.

I recently had a new button show up on my Command Bar. A hand showed up next to my Home Page button. Perhaps it was a new update my computer received, anyway it is now there. Clicking this button will change your mouse pointer from an arrow to a hand when it is on a web page.

The next button shown on the Command bar is the Feeds button. Feeds are fairly new and I will discuss them in the next lesson.

The next button is the Read Mail button. This will let you set up Microsoft Outlook to go out and check your Email accounts for new messages. Setting up Outlook is not hard but you will have to know some details about the settings from your Internet Service Provider. Microsoft Outlook is part of the Microsoft Office package and is not going to be covered in this book.

The next most common command that you will use is the Print command. Clicking the Print button will send the current web page to the printer. The down arrow will also let you see a preview of the printed page.

The Page command has several items associated with it and will be discussed in its own lesson.

The Safety command is also going to be discussed in its own lesson.

There are other options under Tools that we did not discuss. Most of these will be discussed in more detail as we continue to look at the other lessons and toolbars.

The Favorites Toolbar will also have its own lesson.

Isn't it fun having all of this new stuff to learn and play with?

Lesson 6 – 6   Feeds

Feeds were new to me, so I thought they might be new to you. Feeds were introduced as part of Internet Explorer in version 7. They were not something I normally used, so I didn't pay a lot of attention to them. I later found out how great the Feeds were and now I wonder why I didn't use them before.

Feeds, also known as RSS Feeds and syndicated content, are usually used for news and blog websites, where the content of the website is frequently updated. They can also be used to distribute pictures as well as audio and video files.

You will not readily know if the website you are visiting has a feed, so Windows will search the site to see if it has feeds. If it does, the icon for Feeds on the Command Bar will change colors and be available for you to click on. If the website does not have feeds the button is unavailable for you to click on. If you would like to see an example of a web site that has feeds you might want to go to any of the national news websites.

Figure 6-13 shows the Feeds button. If Feeds are available, the icon will turn orange.

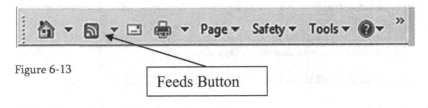

Figure 6-13

Feeds Button

So what do you do if the Feeds button changes color? Remember it changes color to let you know that feeds are available. To view the feeds, you click the down arrow on the right side of the button and then select the feed you want to view from the drop down list.

Oh but wait, there is more to it than just that. You may want to subscribe to some feeds. That does not mean that there is a charge for this service, but I suppose there could be to some feeds. You may see something that looks like Figure 6-14.

You are viewing a feed that contains frequently updated content. When you subscribe to a feed, it is added to the Common Feed List. Updated information from the feed is automatically downloaded to your computer and can be viewed in Internet Explorer and other programs. Learn more about feeds.

✦ Subscribe to this feed

Figure 6-14

By the way, the feeds are still there and available for you to look at even if you don't subscribe. If you click the Subscribe to this feed link, another dialog box will come to my screen and look like Figure 6-15.

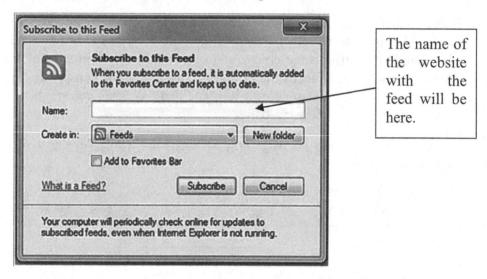

Figure 6-15

Notice what it says in the dialog box!

When you subscribe to a feed, it is automatically added to the Favorites Center and kept up to date. Your computer will periodically check online for updates to subscribed feeds, even when Internet Explorer is not running.

Click on the checkbox for adding this to the Favorites Bar

We will discuss the Favorites bar in an upcoming lesson.

Just think about it for a moment, if I subscribe to this feed my computer will go out and check for new updates periodically and download them to my computer. Anytime I want to see if there are any updates, I can get online and they will be there.

You can also change how often your computer checks for updates to the subscribed sites. We will cover this when we discuss the Favorites Bar.

Lesson 6 – 7   Internet Page Command

The Page command on the Command Bar has increased in the number of items that are displayed with Internet Explorer 8. Why don't we just look at them and see what these commands are all about. Figure 6-16 shows the Page drop down menu.

Click on the Page button

Figure 6-16

The New Window command will open another instance of Internet Explorer. With the Tabs that are available you should not have to use this choice under normal circumstances.

The cut, copy, and paste commands are the same that are in every program form Microsoft on your computer. These should not need any other explanation.

The Blog with Windows Live command will take you to the Windows Live login screen. There will be a lesson about Windows Live a little later. All of the commands about Windows Live will be covered at that time.

The Accelerators command has several things in it and one lesson will be dedicated to Accelerators.

The next section of commands deals with saving the web page and sending this web page to someone as an Email, or sending a link to this webpage to someone via Email. You cannot send either the page or a link to the page from here if you do not have Outlook set up on your computer.

The last one in this section allows you to take a snapshot of the page and put it into Word and then edit the snapshot of the page. You can change the fonts and the pictures or whatever you want. This is the part that you will have trouble believing. If you put a topic in the search textbox of the Word page and click the search button, the Internet Explorer program will show the results of the search. If Internet Explorer is not running, the program will start and then complete the search and show the results. The bad news if you try to save the page as a Word document and reopen it the original pictures and textboxes will not be there. If, however, you save the document as a web page everything will keep working. This is really cool!

The next small section of the drop down menu deals with Compatibility and will be discussed later in the chapter.

The Zoom feature allows you to see the web page in either a larger or smaller size. Figure 6-17 shows the results of clicking on the Zoom command.

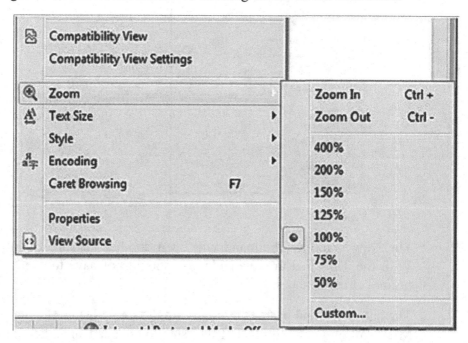

Figure 6-17

By default you should see the contents of the web page at 100% of its original size. This can be changed to a larger or smaller size by clicking on any of the default settings. The Zoom in command will make the contents (text, pictures, and everything) larger by 25% each time you click on it. The Zoom out command will decrease the size of the contents by 25% each time you click on it.

The Text Size command will allow you to change the size of the text on the web page to any of the five provided choices, ranging from largest to smallest.

Back in Chapter two we talked a little about Themes. Web pages have something similar, these are called styles. A lot of the web sites have Cascading Style Sheets. This allows the web designer to have all the pages look similar. If you click on the Style command, you can choose to remove the existing style or continue see the web page with the default style that the web designer used.

Click on Style and then click on the No Style choice

Repeat this only use the default style choice

As you can see, styles make a lot of difference.

If you do not see much of a difference on the website you are on, go to the ez2understandcomputerbooks.com website and then try it again.

The Encoding is something you should normally never have to adjust. You can click on the different choices that are available and see the differences they make, but change it back to the original settings when you are finished.

The Caret Browsing will put a movable cursor on the web page so you can use the keyboard to select text and navigate around a web page instead of using the mouse. You will be able to use the arrow keys as well as the Home, End, Page Up, and Page Down keys. This is not as easy as using the mouse and unless you just don't like the mouse I wouldn't recommend turning this on.

The Properties command will give you some basic information about the web page. As you can see in Figure 6-18, there is not a lot of information that the average user will want.

Figure 6-18

The last command in the drop down menu is the View Source command. This will bring the actual source code the web designer used to construct the web site to the screen for you to look at and examine. Most of this will mean nothing to the average user and it will just look confusing.

This will take us to the next lesson: The Safety menu.

If you have the Page Menu open, click anywhere outside of the menu to close it

Lesson 6 – 8   Internet Safety Command

The Safety command (button) deals with using the provided safety features in Internet Explorer 8. Figure 6-19 shows the drop down menu if you click the Safety button (or drop down arrow).

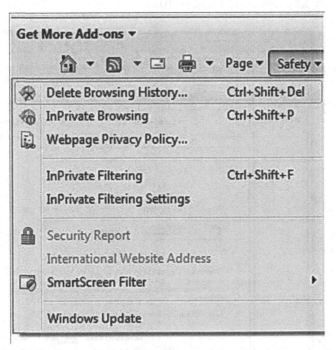

Figure 6-19

The first thing that you will notice is that you can delete your browsing history form here. The last time we did this we used the Tools command to delete the history. If you click on the Delete Browsing History command you will get the same dialog box that was shown in Figure 6-8. The choices are still the same as they were then. You might also want to notice the keyboard shortcut that is available.

The next command deals with InPrivate Browsing.

When you are on the internet, Internet Explorer stores data about what web sites you visited. This includes such things as your browsing history, temporary internet files and cookies from the web pages that you visited. Before you get too mad, you might as well know that all web browsers do this.

If you do not like this, you can turn on the "InPrivate browsing" feature. This is new and was not available before Internet Explorer 8. This has been standard on Safari, which is the internet browser on Mac, for some time. Microsoft is finally catching up on some of these things.

189

This is not a onetime catch all command. You have to turn this on every time you want to use it. It is that way on Mac also. Not every web browser has this feature and you need to consider this if you decide to try one of the other web browsers on the market.

To turn In Private browsing on, you need to click the "InPrivate" command from the drop down menu. If you click on this command you will get a new browsing window as shown in Figure 6-20.

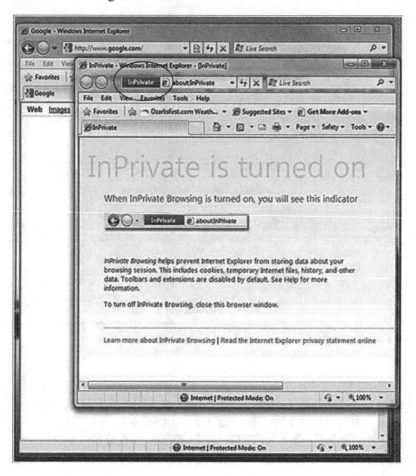

Figure 6-20

The original browsing window is still there and still working. You can see the first Internet Explorer window behind the private browsing window.

The "InPrivate" browsing will only be active when you use this window (the one that has "InPrivate" in the address box) to get to web sites. You can have more than one tab open and each tab will also have private browsing. Remember if you close this window or use the first window the privacy will not be turned on.

Having the private browsing turned on will prevent Internet Explorer from storing the information about your browsing session. This will also stop anyone else from checking the web sites that you have visited.

To turn the private browsing off, simply close the window.

The next item down on the menu is the Webpage privacy policy. This section deals with cookies. Cookies, although I really like peanut butter cookies, are not always good. Cookies are something that websites use to gather information about website usage. Some cookies are not necessarily bad. If there is a web site that you visit often, a cookie may allow you to go to the website and not have to login every time. There are also cookies that are not so good. These cookies will track the websites you visit and could put your privacy at risk.

The next question is should I allow or block cookies? With Internet Explorer you can decide if cookies are allowed to be stored on your computer. You can also allow some cookies and block other cookies. Your first response will probably be that you will just block all cookies and never have to worry about it. If you do that you may not be able to get on some websites, or the website may not load correctly. Let's talk about the cookies you will want to block. Then we will talk about how to block them.

Since we cannot list every possible website that we do not want cookies from, we can specify the general type of cookies that we do not want.

Click the mouse on Website Private Policy

The following dialog box will come to the screen (See Figure 6-21).

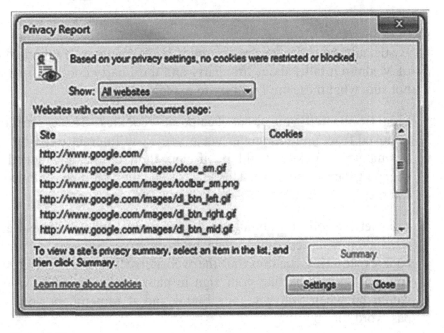

Figure 6-21

Click on the Settings button

Another dialog box, the Internet Options Dialog box, will appear on your screen. We discussed this before and you have seen this dialog box before. Figure 6-22 shows the Internet Options Dialog Box.

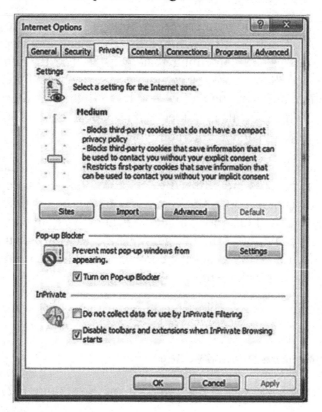

Figure 6-22

By default the slider toward the top is set to medium. If you read the description of Medium it talks about first party and third party cookies. Just in case you are not sure what these are I will try to explain.

First let me tell you about Temporary cookies. These are cookies that are removed from your computer when you close Internet Explorer. An example of a temporary cookie would be if you shop online and added items to your shopping cart to purchase. This kind of cookie would be removed when you turn Internet Explorer off.

Next let me tell you about saved (persistent) cookies. These are cookies that remain on your computer even after Internet Explorer is turned off. These cookies store information that the website will need if you go back later. This could be something like your sign in name and password. An example of this would be when you get your Email and it remembers your user name and password for you.

First party cookies can be either temporary or saved. These cookies come from the website you are visiting. They will reuse this information the next time you visit their website.

Third party cookies are cookies do not come from the website you visited, but are put there by companies that advertise on the website you visited. These cookies can track the websites you visit and that information can be sold or used to help the advertiser with their marketing strategies.

Earlier I told you that I thought it would be better if you changed the setting from medium to medium high. I don't think that I would go below this setting. You might even want to go up to high. If you do go to high and you cannot get to some of the websites you want to visit, you may need to come back and change this to medium high.

Bet you didn't plan on learning that much about cookies did you?

Next we should talk about the "InPrivate Filtering" feature."

The first time you click on the "InPrivate Filtering" choice from the drop down menu under Safety, the dialog box shown in Figure 6-23 will jump onto your screen. After the first time, clicking the "InPrivate Filtering" command will toggle the feature on and off.

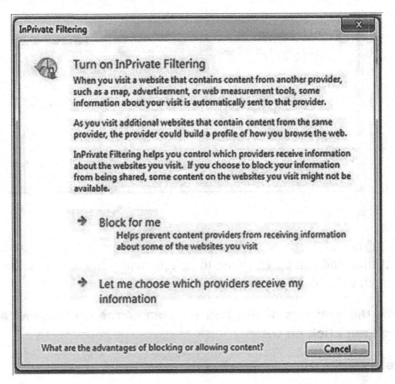

Figure 6-23

If you visit a website and there is an advertiser on it, your information could automatically be sent to the advertiser. If you visit another website that information could also be sent to the advertiser. Then a profile could be built about you and how you browse the internet. You could also start receiving Emails from these companies advertising their products. Internet Explorer allows you to turn on filtering to limit how much information can be provided. The two choices are let Internet Explorer do the blocking for you or you can choose which providers will receive the information. It will be easier to let Internet Explorer block for you.

If you want to make the choices for yourself and you click on the bottom choice the following dialog box will come to the screen.

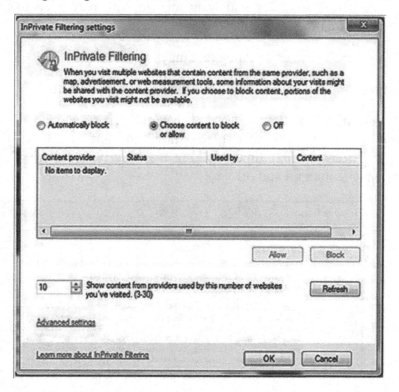

Figure 6-24

If you start blocking these manually, you will probably get tired of it in a very short time and end up choosing to automatically block, so we might just as well let Internet Explorer do it for us.`

If you have the settings dialog box on your screen, click on "Automatically Block" and then click on OK

Otherwise

Click on block for me

194

Next, let's talk about the SmartScreen Filter.

The SmartScreen Filter is a tool (feature) in Internet Explorer 8 that helps detect phishing websites. The easiest way to describe phishing is to remind you of some scams that were (and probably still are) going around to get you to give your personal information, such as social security number or bank account numbers to someone posing as a government official or someone from your bank. This is all done over the internet, and they can be very convincing. Sometimes this will come in the form of an Email and look very official. The information, if you provide it, is directed to a fraudulent website and then possibly used for identity theft.

The "SmartScreen Filter" will work silently in the background while you are browsing websites; analyzing the websites you visit. You won't even know it is there unless it finds something it thinks is suspicious. If it finds something, you will get a message box telling you that there is the possibility of fraudulent activity and advising you to proceed with caution.

SmartScreen also checks the websites you visit against a list of known phishing websites. SmartScreen also check for malicious software. This includes viruses and worms. SmartScreen will also check the website you visit against a list of known websites that have malicious software and will block any suspected malicious software from being installed.

The last thing on the Safety command is the Windows Update feature. You can click on this at any time to see if there are any updates available for your computer.

The only thing on the command bar that we have not really talked about is the Help button. The help button is the small blue question mark with the circle around it and is located on the far right end of the command bar. Back in the olden days a series of books came with your computer and the software when you purchased it, on how to use the product. This was expensive and cumbersome so the manufactures started putting everything on CDs. After a while the manufactures just started including the manuals as part of the software in a section called Help. There is more information included in the help section than the average user will ever need, but it is all there. You can click your mouse on the question mark and type any subject about Windows or any of its programs, in the search box and Windows will display any information it has on the subject. It took awhile, but the help section is getting easier to understand. At one time you felt like you had to be a computer geek to understand what they were talking about when you tried to read something in the help section. It is getting better.

I think by now you probably need a break.

Close Internet Explorer

Lesson 6 - 9   Search Engines

In this lesson we will discuss how to find a website by using a search engine. A search engine is a tool that is provided by certain web sites that will let you search the internet for various topics. One popular search engine is Google. In this lesson we will look at a couple of the common search engines available. I cannot attempt to list all of the search engines and I do not make any attempt to recommend one search engine over another one. I will give you a brief description of how to use a search engine and then we will look at the Live Search feature in Internet Explorer 8.

Open Internet Explorer if it is not open

In the address box type google.com and press the Enter key

If you prefer a different search engine you can type that in instead of Google. Either way there will be a textbox for you to type in. In this box you can type just about anything you want to search for and the search engine will find the websites that have that on it. The search box will look similar to Figure 6-25.

Figure 6-25

Type something like Race Horses **in the textbox and then click on the search button**

The screen should change in a few seconds and show you the results. My search for race horses turned up 4,590,000 results. If you want to go to one of the websites that has something to do with race horses, click on the underlined name and you will be whisked away to the website.

This is basically how search engines work.

Microsoft has made this process a little easier. If you want to search the internet for a topic or website, you don't have to go to a different web page to find a search engine. Internet Explorer has the Live Search feature always available no matter what web page you are visiting. The Live Search area is on the upper right section of the Internet Explorer screen and is shown in Figure 6-26.

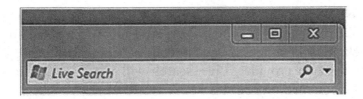

Figure 6-26

This works the same way that any other search engine works. You type in what you want to search for and either press the Enter key or click on the magnifying glass in the right corner.

If you want to choose the search engine you would like to use when you use the Live Search feature, you can click the down arrow on the right end of the search bar. Figure 6-27 shows the drop down menu.

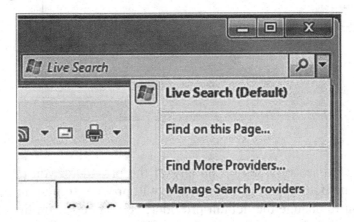

Figure 6-27

Click the down arrow and then click on Find More Providers

You will now have the Add-ons Gallery: Search Providers page showing on your screen. This is a list of search providers that can be added to Internet Explorer as an add-on. If you see a provider that you like, you can add this to Internet Explorer by clicking on the Add to Internet Explorer button. If you do add this, another dialog box will come to the screen (See Figure 6-28).

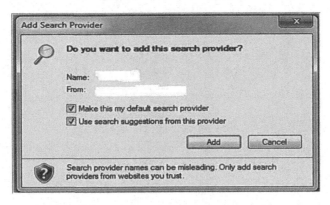

Figure 6-28

The "Name" section will show the name of the provider and the "From" section will show the company that provided the Add-on. To finish adding the provider, simply click the Add button. If you click the checkbox making this new provider your default search provider, Live Search will use it as the first choice when searching the internet.

If you would like to see the search providers being used on your computer, click the Manage Search Providers choice on the drop down menu. The dialog box shown in Figure 6-29 will come onto the screen.

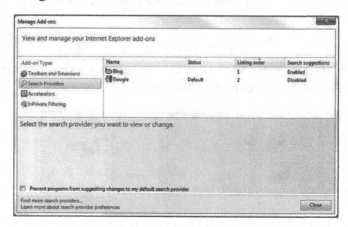

Figure 6-29

If you want to see information on any of the providers, you can click on their name and the bottom part of the window will provide some information for you. If you want to change the default provider, you can click on the one you would like to be the default provider and then click on Set as Default. You can also choose to remove one of the providers if you click on it and then click the Remove button. You can also prevent programs from suggesting changes to your default search provider by clicking the appropriate checkbox. That is probably enough about search engines; let's move on to something else.

Click on the Home command and go back to your home page

Lesson 6-10   Favorites

Okay, now you have found a few websites that you like and you wanted to make sure you could find then again so you probably wrote down the address so that you could find the site again. In this lesson I am going to show you an easier way to go back to the website, and you won't even have to write the address down.

In lesson 6-2 we learned how to get to a web page; actually we learned how to get to two web pages: Google and the Branson United Methodist Church web page. Let's suppose that you like church music and want to go to one of the sites that offer hymns that you can listen to. You will want to go to go to this site often and need a way to save the web page so you can get to it.

Go to the website www.hymnsite.com

This is a website that allows you to listen to music and even download some of the music if you want. You can also search for different hymns and listen to the ones that are available. We will want to save this web page in our Favorites list.

The Favorites button is in the upper left corner just below the Back arrow (See Figure 6-30).

Figure 6-30

Click the Favorites button

The following menu will drop down.

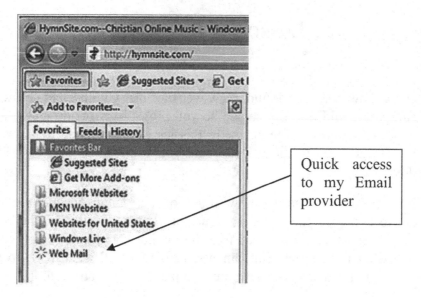

Quick access to my Email provider

Figure 6-31

There are two ways you can add this to your favorites. You can click on the Add to Favorites button and the Add a Favorite Dialog box will come to the screen as shown in Figure 6-32.

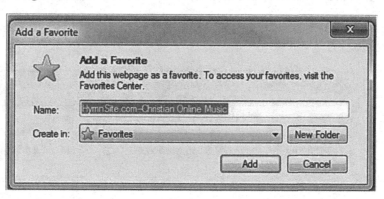

Figure 6-32

When you click the Add button the link will then be added just like the link to my Email provider. You can also shorten the name if you don't want the entire name to be on the list. In this case you would probably shorten it to Hymn Site.

Click the Add button and add this to your favorites

The second way is to click the down arrow at the end of the Add to Favorites button. If you click the down arrow, the following menu will drop down.

Figure 6-33

Clicking the Add to Favorites Bar will add the link to the bar, right next to "Suggested Sites", and it will be visible all of the time.

If you have a web page in the favorites and you want it to be always visible in the Favorites Bar you can right-click on the link in the favorites and choose Add to Favorites Bar from the shortcut menu (See Figure 6-34).

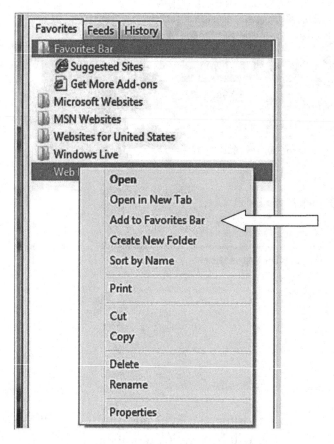

Figure 6-34

Adding a website to your favorites was so easy that I am going to stray for just a moment.

Some websites that were designed with older browsers may not look the very best that they could look with Internet Explorer 8. To help fix this, Internet Explorer has the Compatibility View option. Using this may not improve every website that you visit, but it could help on some of them. Using the Compatibility view is very easy; all you have to do is click on the icon on the Address Bar. Figure 6-35 shows the Compatibility icon.

Figure 6-35

Click on the Compatibility button and see if it makes a difference to the browser view of the website

Earlier when we talked about Feeds, I said we would discuss them a little more when we talked about Favorites. Well now is the time.

Click on the Favorites button again to bring the drop down menu to the screen

There is a tab on the top of the menu that has Feeds on it.

Click on the Feeds tab

Figure 6-36 shows the Feeds tab.

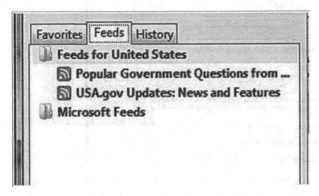

Figure 6-36

Any feeds that you have subscribed to are going to be listed in this section. You can view the feeds by clicking on them with the mouse. If you want the feeds to be displayed on the favorites bar you will need to right-click on the link and then choose Add to Favorites Bar from the shortcut menu.

I also told you that you could change how often the feeds would be updated. We will see how to change the default setting next.

Right-click on one of the Feeds

A new dialog box will come to the screen and is shown in Figure 6-37.

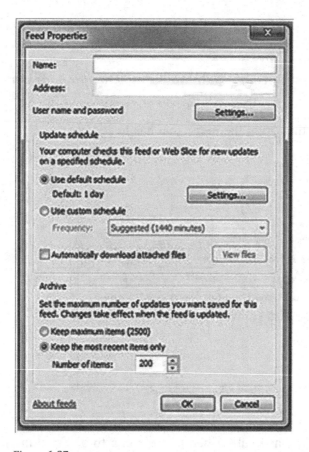

Figure 6-37

By default, your computer will only check the site for updates once a day. If this is not often enough you can click the radio button next to Use Custom Schedule and choose how often it updates from the listed choices. These choices range from every 15 minutes to never.

When you have made your choice, click the OK button and close Internet Explorer

Lesson 6 – 11   Printing Web Pages

This will be a very short lesson, because there is not much to printing a web page. The web page may have only text on it and you need a copy of the text for your research or paper or whatever. The web page may have pictures on it that you like and want to print them out. No matter what the case, you should be concerned with copyright laws. Everything that is on the internet is covered by copyright laws. You cannot just print something. You need permission to reprint something you find on the internet.

Having said that, if you have permission and want to print the contents of a web page all you have to do is click the print button, which you will find on the Command bar next to the tabs.

If you have permission and want to download a picture that you found on the internet, you can right-click on the picture and choose "Save Picture As" from the shortcut menu. Figure 6-38 shows the shortcut menu.

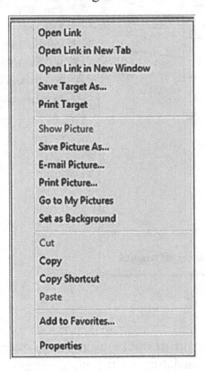

Figure 6-38

As you can see, there are several choices that you can make from here, such as printing.

Do not forget, the things you find on the internet belong to someone; you need permission to copy them.

Lesson 6 – 12   Accelerators

Accelerators are fairly new and you may not know what they are and how to use them, so we will go over the basics in this lesson.

You use an Accelerator with text that you have selected on a website. Let's suppose that you were on a website that had the word Angels on it, and you wanted some additional information on the word Angels. You would select the word (you could double-click on the word or use the mouse to select it) and let your mouse hover over the word for a second or two until the Accelerators icon became visible. You would then click on the icon and the list of available Accelerators would become visible. Figure 6-39 shows the various parts of the Accelerator menu.

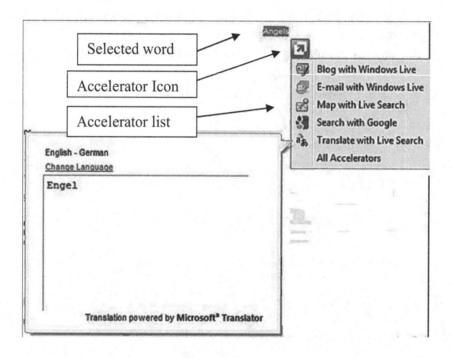

Figure 6-39

The above example shows the result of clicking on the Translate choice. This shows the English to German translation of the word Angel. To translate to another language you just need to click on Change language and choose from the list of available languages.

If the text was an address, you could click on the Map choice and see where the address was located. If you wanted to start a discussion on this subject, you would click on the blog choice and be taken to the Windows Live Blog site where you start a discussion about the subject.

What I noticed was there was no dictionary in my list of Accelerators. What if I was not sure what the selected word meant? I think I need to add a dictionary to my Accelerator list.

If I move my mouse to the "All Accelerators" choice another menu will slide out to the side. From here I can find more Accelerators (See Figure 6-40).

Figure 6-40

When I click on the "Find More Accelerators" choice a new tab will open. The new tab will show the Add-on Gallery and have a list of all of the Accelerators that are available. On the left side you can see the link to various types of Accelerators available. Figure 6-41 shows a partial screen shot of this.

Figure 6-41

To find a dictionary, you would click on Dictionaries and Reference. From here you can choose one of the several choices that are listed. You might want to check the rating of the Accelerator before you add it to Internet Explorer. I like using the "Define with Live Search" Accelerator.

There are several categories to choose from and several Accelerators in each category. Look around, play, and have some fun.

Some of the Accelerators have a preview of the information that is available. This will often provide you with the information you need. The ones using the Live Search seem to have a preview more than the others. If there is no preview, you can click on the Accelerator and you will be taken to the service for that Accelerator.

This is something really cool. You can copy text to the Clipboard, and then open Internet Explorer. You can then click on a new tab and on the right side about half way down click on Use an Accelerator. The last text you copied will be listed under "Use this text" and then you can choose one of the Accelerators to define, or map an address, or translate to another language.

Lesson 6 – 13   Virus Protection

I am sure you have listened to the news and heard reports whenever a new computer virus starts infecting computers. Most viruses seem to come from unsolicited Emails which contain the virus or from downloading something from the internet. You can't stop using the internet or stop getting Emails, so what can you do?

A good anti-virus program is worth its weight in gold. An anti-virus program will scan your computer at scheduled intervals and check for any known viruses and threats. The anti-virus program will also constantly update itself so it can identify threats as they are found on the internet.

There are many commercial products available and one probably came with your computer when you purchased it. It probably had a trial period that you could try it out and then pay for the program after the trial period.

There are also free versions available on the internet. I would not even attempt to list the products that are available or recommend one product over another. If you want to find out about a product, ask one of your friends which anti-virus program they use and if they are satisfied with it. The odds are that they will know which one they are using. You can also read about the different anti-viruses in computer magazines.

Regardless of which one you choose, having one is vital to the welfare of your computer.

Microsoft started including a program called Windows Defender a while back. This program is designed to search for Spyware. Spyware is software that can install itself on your computer without your consent. Some of these, called malicious software, are designed to slow your computer down and affect the way your computer runs. Every time you start your computer Windows Defender will start running in the background (if it is turned on, and I believe it is turned on by default).

Windows Defender can help you in two ways. First it offers real-time protection. This means that you will be alerted any time spyware attempts to install itself on your computer or it attempts to change any of Windows important settings. Windows Defender will also scan your computer for any spyware that is installed on your computer.

By default Windows Defender will scan your computer every day at approximately 2:00 a.m.

You may want to check the settings on a few of the options, so let's look at them.

Open the Control Panel and then click on Windows Defender

The easiest way to find the Windows Defender is to make sure you are viewing All Control Panel Items. The items are in alphabetical order, so Windows Defender will be toward the end.

You should see the following dialog box on the screen when you open Windows Defender.

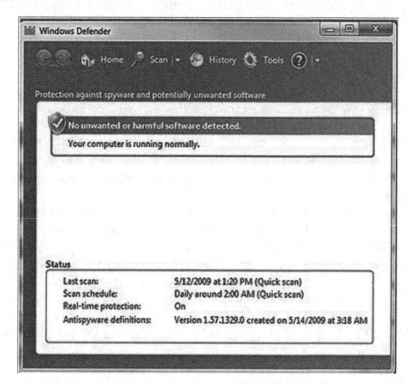

Figure 6-42

Hopefully your box will say the same thing mine says; that your computer is running normally and there was no harmful software found. It should also tell you the last time your computer was scanned. If your computer has never been scanned, because you probably turn it off every night, you may want to click the down arrow next to scan and then click on Quick Scan. This will start the Windows Defender scanning process. You can keep working while the scan is running, but your computer may run a little slower than normal.

Click on the Tools command at the top and then click on Options

The Automatic Scanning choices will be showing on the screen, and you should let it scan automatically. It would be best to let it scan every day. You can set the time to whatever is convenient for you. Normally the Quick Scan will work just fine, but you may want to run a full scan every once in awhile.

Make sure the checkbox next to Check for Update Definitions before scanning is checked

This will let Windows Defender check the master list of known spyware and malicious software for anything new before it scans. This is a good thing. The part about scanning only when the system is idle can go either way, so it doesn't matter. Like I said before the only downside to letting it scan while you are working on the computer is that the computer will run a little slower.

Click on the "Real Time Protection" command on the left side and make sure all three checkboxes are checked

Click on the "Advanced" command and make sure all of the checkboxes are checked

Click on the "Administrator" command and make sure both checkboxes are checked

Save all of the changes

Close the Windows Defender dialog box

The last thing we will talk about is your Firewall.

Microsoft includes a Firewall as part of the operating system. A Firewall is a software program (although it could also be a piece of hardware) that checks all of the information that comes into your computer from the internet or your private network, such as at work. The Firewall then either allows the incoming information to pass through it and enter your computer or it blocks it and prevents it from entering your computer. The Firewall can help prevent hacker and malicious software (such as worms) from getting access to your computer. The Firewall can also stop your computer from sending malicious software to other people's computers.

In the Control Panel click on Windows Firewall

Your screen will change to show the Firewall Dialog box as shown in Figure 6-43.

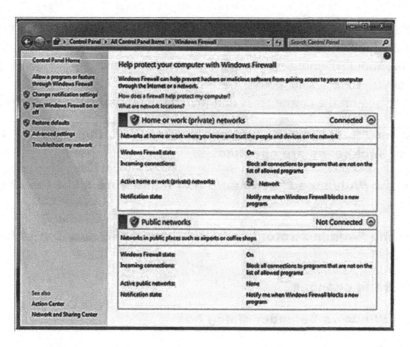

Figure 6-43

Basically you want to make sure this is turned on. If it is not, use the command on the left side to turn it on.

Close the Control Panel

Chapter Seven   Email

Electronic Mail (Email) is becoming part of everyone's life. In fact, it amazes me that the Post Office is still in business. With Email you can almost instantaneously send and receive mail. Not everyone feels comfortable using Email, and some people do not use it at all, so we might just as well explain it.

Lesson 7 – 1   Sending an Email

Every Internet Service Provider (ISP) is a little different in the way the screen looks. Knowing this I will try to describe generalities. The concept is the same no matter which service you use. So let's see what happens.

Get online and go to the place where you receive your Email and login with your user name and password

This could be your local cable or telephone company home page or it could be AOL or Hot Mail or G Mail or any of hundreds of other providers. Somewhere on the screen will be a button that has Compose or Create New Mail or something similar on it.

Click on the button to compose or create a new Email

This will bring up a new screen. The screen should look similar to Figure 7-1.

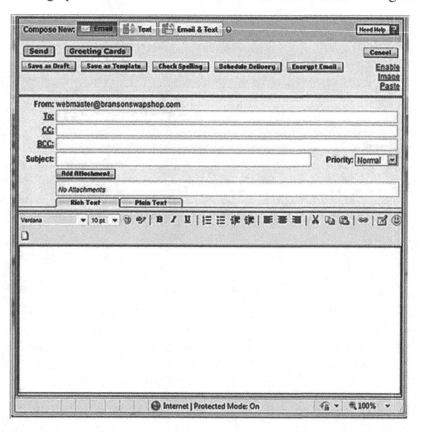

Figure 7-1

Let me repeat; your screen will probably look different than this one, but it will have the same general things on it. Let's take a look at what each of the sections is used for.

The "To" section is where you type the Email address of the person who is going to receive the Email. Think of this in this manor, if you were writing a regular letter to someone you would need to put the person's name and address on the envelope for them to get the letter. In an Email we need to put the person's electronic address in the "To" section for them to get the Email.

The "CC" refers to Carbon Copy. If you are old enough to remember way back when we used carbon paper to make a copy of a letter, you will understand what a carbon copy is. If you are not old enough to remember carbon paper, you probably already know what "CC" stands for. In either case if you put someone else's address here they will receive a copy of the Email. If you have ever received an Email with about a dozen addresses at the top, this is why; they received a copy of the Email.

The "Bcc" (Blind Carbon Copy) is similar to the "CC" with one exception; the person receiving the copy of the Email will not have their address show up anywhere on the Email.

The "Subject" line is where you type a quick indication of what the Email is all about. You would be surprised how many Emails get sent to the trash can because someone looked at the subject line and then decided not to read the Email. It is very easy to hit the Delete key, so make sure the subject line has the reason you sent the Email.

The large area below this is where you type the contents of the Email. This is the important part. This is why you sent the Email, to get this information to the person.

Now all you have to do is click on the "Send" button and the Email will whisk off into cyberspace and wait for them to open it.

Lesson 7 – 2   Attachments

Well sending an Email was pretty easy, but what if you want to send a picture or a file?

Go back to your Email provider and compose another Email

Somewhere on the screen will be a place for you to add an attachment. An attachment is something that you did not include in the body of the message. It could have been a long letter that you have already typed in Word or WordPad and you just did not want to type it again, or it could be a picture of the new baby that you want the person receiving the Email to see. There should be a button that has "Attach" on it, or it could say "Add Attachment", or it could just look like a paperclip. Clicking on this will allow you to add the attachment. If you click on this, you should get a screen similar to either Figure 7-2, or perhaps 7-3.

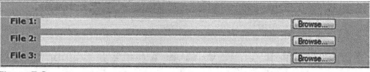

Figure 7-2

Figure 7-3

If your screen looks similar to Figure 7-2, you will need to click the Browse button to look for your file on the computer. Usually this will be one of your document or pictures. Clicking on the Browse button will take you to the screen shown in Figure 7-3.

Usually you will click on Documents or Pictures and find the file you want to send and then click the Open button. You may have to click one more OK button to go back to the screen where you typed in the address of the person getting the Email and then click the "Send" button.

Lesson 7 – 3   Reading Emails

There is probably not much new in this lesson, but we will look at reading an Email and viewing and opening an Attachment. I have had people tell me that they get picture all the time of their grandchildren and families and have never seen one of the pictures. That is why I am including this lesson.

When you are checking your Email you may have more than one Email to look at. I am including two different examples of the way an Email may look.

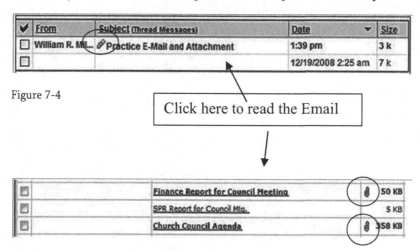

Figure 7-4

Figure 7-5

Since these are real Emails, I erased the name and address of the sender.

In the above examples we have Emails with an attachment and without an attachment. You can spot the ones with an attachment because they have the paperclip somewhere on the line that shows the Email. To read an Email I need to click somewhere on the Subject part of the message.

When I click on the subject line, I will get a screen that looks like the following two Figures (one for each ISP).

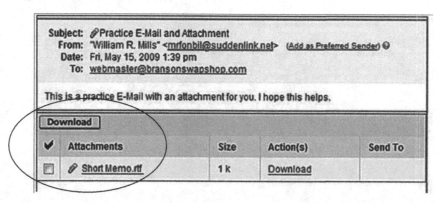

Figure 7-6

218

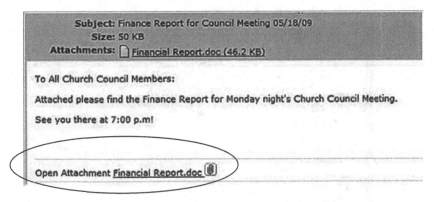

Figure 7-7

To see the attachment you might have to click on it (sometimes pictures will automatically open and be seen below the actual Email message). When you click on the attachment, you will more than likely get a dialog box similar to the one shown in Figure 7-8.

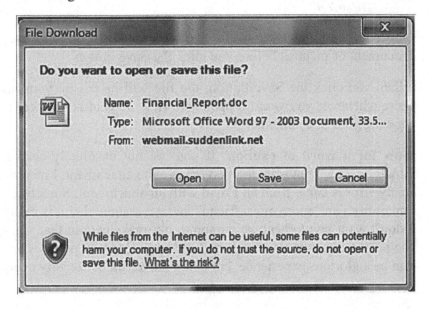

Figure 7-8

From here there are a couple of things that you can do, besides cancelling your actions, you can open the file and look at the contents or you can save the file to your hard drive. If it is something important, such as a picture of your family member, you will probably want to save it on the computer. If it not something really important, you may just want to open it and look at it. If you decide to save the file and you click on the Save button, you will get a screen like Figure 7-9.

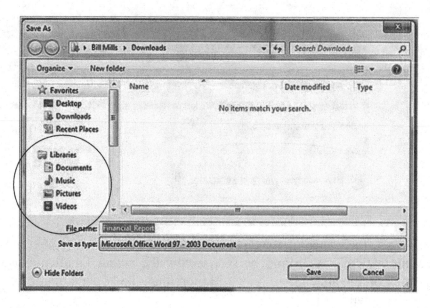

Figure 7-9

Depending on the type of file you are saving you may want to click on Document or pictures before you click the Save button.

When you click the Save button, the file will be put on your hard disk and be there whenever you want to open and view it. That is all there is to saving and viewing your attachments.

Now for a word of caution: If you are not absolutely certain who sent the Email, you may not want to open or save the attachment. I mentioned earlier that many viruses come from an Email with an attachment. Sometimes we receive an Email from an "anonymous friend" and we assume that it is really from a friend. I don't want you to become paranoid, but make certain that you know and trust the person that sent you the Email. Receiving family pictures and stuff like that can be a glorious experience, but being naïve can be a costly experience.

Lesson 7 – 4   Replying & Forwarding

There may come a time when you want to respond to an Email and you don't want to go back to the Compose button and then type in the address of the person you want to send the response to. Just about everyone thought that would be a real drag, so they provided an easier way to respond to an Email.

While you are reading the Email you will find (usually up by the top) a button that has "Reply" on it. An example of this is shown in Figure 7-10.

Figure 7-10

Clicking on the Reply button will bring the Compose (or Create) window back to the screen and all of the essential parts will already be filled in for you. All you have to do is type what you want in the main body part and click the Send button. That is all there is to Replying to an Email.

Sometimes you will get an Email that is so great that you will want to send it to someone else. You can do that also. Instead of clicking the Reply button, click the Forward button. This will also bring the Compose window to the screen with everything filled in except the address of the person that you want to receive this great Email. All you have to do then is fill the "To" section in and click the send button.

It is getting difficult to find something that is hard to do.

Lesson 7 – 5   Address Books

A frequent question that comes up is: How can I remember someone's Email address? There is an easier way than trying to write them all down and keep them in a file somewhere. Most Internet Service Providers provide to you an Address Book where you can store contact information.

Go back to your Email provider like you were going to check your Emails

Look around on the screen and see if you see a button that has something like Address Book on it. It might look something like Figure 7-11.

Figure 7-11

When you find it, click on the button

Your window should now have a new screen on it and it might look similar to Figure 7-12.

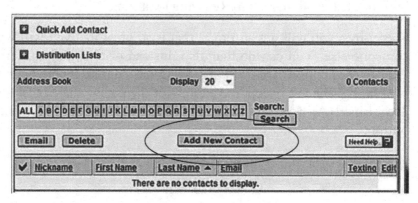

Figure 7-12

Somewhere on the screen will be a button that has something like Add New Contact on it or it may just say New Contact.

Click the button to create a new contact

The screen will change again and have a new window on it. It should look similar to Figure 7-13.

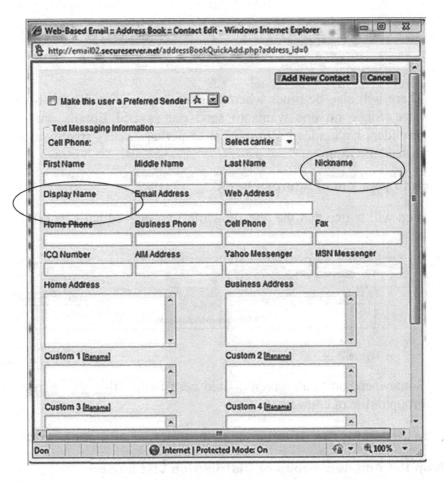

Figure 7-13

To send Emails all you need is the Display Name and Email Address. In some cases you might have to put in the Nickname for the name to show in the address book.

Add a contact to the Address Book and then click Add Contact or Save Contact

After you have added a contact you can edit the information by clicking on the Edit or Edit Contact button (usually located somewhere on the line where the name is). If you do not have an Edit button you might be able to click on the contact name and that will allow you to edit the information.

To send an Email using the Address Book, all you should have to do is click on the Email address and this will bring the Send Email window to the screen with the "To" section already filled out with the person's Email address.

Lesson 7 – 6 Group Emails

There will also be times when you need to send an Email to several people at once. Since no one wants to send out several Emails one at a time, most providers have added a group list or perhaps yours might say Distribution list instead of Group.

Go back to your Email address book

You will notice that the one shown in Figure 7-14 calls its group a Distribution List.

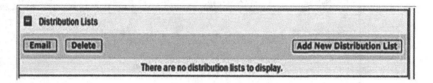

Figure 7-14

Somewhere on your screen should be a button that will allow you to make a group or list of contacts.

Find the button for the Group or Distribution List and click on it

Click on the Add new Group or Distribution List button

A new window should come to the screen and look similar to Figure 7-15.

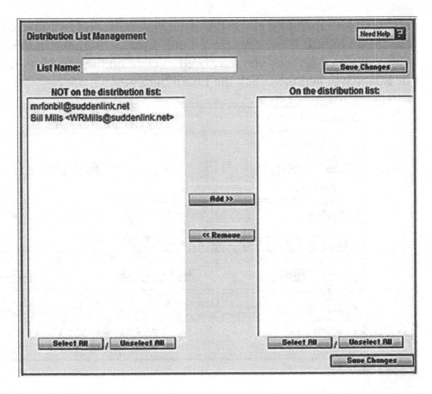

Figure 7-15

All you have to do at this point is give your group (or list) a name and then add contacts into it.

For our practice, let's use a name like Close Friends.

Type Close Friends **in the Name section**

To add someone to the group, we need to click on their name on the left side (the one that says they are not in the group) and then click the Add button in the center. This will add this contact to the list.

Click on one of your contacts and then using the Add button, add them to the group

When you are finished adding contacts all you have to do is click the Save button.

When you are finished adding contacts, click the Save button

Depending on your Internet Service Provider your list may show up in the groups (or Distribution List) section or the name of the group may just appear as any other contact.

You send an Email to a group the same way you send an Email to an individual. The ISP in Figure 7-16 keeps the groups separate from the individual contacts. If your ISP also does this, you may need to click on the checkbox next to the group (or List) name and then click the Email button.

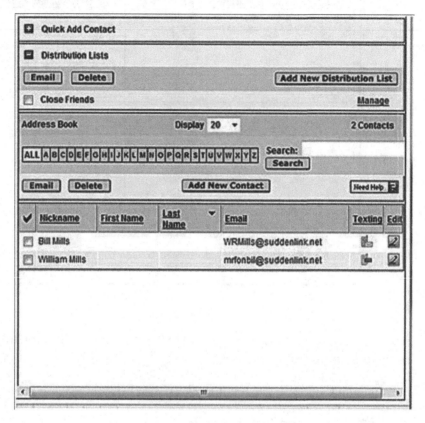

Figure 7-16

Close Internet Explorer when you are finished

Lesson 7 – 7   Virus Scanning

This lesson is just some advice for you, but you should read it carefully. Most Anti-virus programs come with a feature that asks if you want to scan all Emails for viruses. This is a good feature. If you activate it all incoming and outgoing messages will be scanned before they enter or leave your computer to make sure they do not contain any known viruses. This is something that will protect you and the people you send Emails to.

I mentioned before that one source of viruses that can infect your computer come inside of an Email. I personally know several people who have opened attachments (or Greeting Cards from an unsolicited "friend") only to have their computer lock up or crash a little later.

Let your Anti-virus software scan your Email before you open it. Also make sure your Anti-virus software is allowed to update itself on a daily basis so it can keep up with all of the known viruses.

You may think that scanning your outgoing Email is not necessary because you would not send malicious software to anyone. Some viruses have the ability to go to your address book and start sending out infected messages to your friends and family. The person on the other end will think that the Email is from you, so they will open it. You need to protect yourself and your friends.

This was short, but a very important lesson.

Chapter Eight   Desktop Gadgets

If you have used Windows Vista, you may have loved having all of the Gadgets that were available for the Sidebar. You may not have loved the Sidebar, just the Gadgets that were on them. The Sidebar was an area of the screen that was set aside for the Gadgets and could be hidden or displayed as you desired. Not everyone liked having this area visible on the screen. I turned mine off because my monitor was not that large and it seemed to take up valuable screen space. I did, however, like having the Gadgets on my monitor.

Gadgets for the people who are not familiar with the name are "Mini-programs" that you can have on your Desktop. I usually have the clock, CPU meter, and local weather on my monitor screen. One of the great things about Desktop Gadgets is that you can move them around on the screen and there is no designated area set aside for them. Since they are running on the Desktop, any programs you open will cover the Gadget up and you don't lose part of the screen area.

In this chapter we will discuss adding and removing Gadgets and how to get more Gadgets online.

Lesson 8 – 1   Adding Gadgets

Windows 7 will come with some Gadgets already on your computer and ready for you to add them to your desktop. As you will see, adding Gadgets is so easy you will wonder why we even had a lesson on it.

Click the Start button

Click All Programs

The Desktop Gadget Gallery is listed toward the top of the "All Programs" menu (See Figure 8-1).

Figure 8-1

Click on Desktop Gadget Gallery

A gallery of the available Desktop Gadgets will spring onto your computer screen (See Figure 8-2). This gallery only shows the Gadgets that came with your computer. There are a lot more Gadgets available as you will see a little later.

Figure 8-2

Now that you can see the Gadgets that are on your computer, you will probably want to add some of them to your Desktop. To add a Gadget to your Desktop all you have to do is double-click on it.

Double-click on the Clock Gadget

The Clock Gadget will jump onto your monitor, in the upper right hand corner of the screen. This is shown in Figure 8-3.

Figure 8-3

This may not be the image of a clock that you prefer to see. Well fret not; there are other images you can use if you prefer them.

Move your mouse over to the Clock Gadget

A menu will appear on the right side of the Gadget. There are a few basic things that you can do from here. The "X" at the top will let you close the Gadget. The wrench in the middle will let you choose from the available options. The small dots at the bottom will let you drag the Gadget to a new location on the Desktop. Figure 8-4 shows these.

Figure 8-4

Click the Options choice

Hey that looks like another dialog box (See Figure 8-5).

Figure 8-5

As you can see from the dialog box this is only the first view available of the clock. There are seven other choices for the clock that will appear on the screen.

Using the arrows look at the other choices available

You may prefer one of the other choices for the one you will look at every time you look at your computer screen. If you find one that you like better, leave that choice showing on the screen. Personally, I prefer to also see the second hand on my clock. If you prefer to also see the second hand make sure the checkbox next to "Show the second hand" is checked.

When you are finished, click the OK button

The Gadget will now show on your screen until you decide to remove it.

Removing a Desktop Gadget is as easy as adding one. To remove a Desktop Gadget move your mouse over to the Gadget and click on the "X" on the menu.

Let's not forget about the shortcut menu that will appear if you right-click on the Gadget. Figure 8-6 shows the shortcut menu.

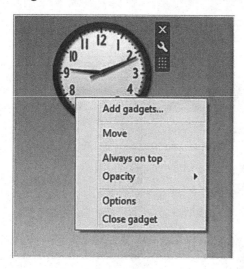

Figure 8-6

The Add Gadgets menu choice will take you to the screen that was used to add the Gadgets. The Options choice will take you to the Options that we just finished using. The Close choice is self-explanatory.

Earlier I mentioned that the Gadgets would be covered up by an open program and that is true unless you choose the Always on top choice. This will leave the Desktop Gadget visible at all times, and the open windows will be behind the Gadget.

The Opacity choice will determine how faded or washed out the Gadget will appear. You can change the Opacity so that the Gadget is almost entirely see-through or have it so that nothing can be seen behind it.

Change the Always on top and the Opacity options to see what affect they have on the Gadget

Add whatever Desktop Gadgets that you might like to have on your screen

Lesson 8 – 2   Gadgets Online

If you want more Desktop Gadgets than are available on your computer, you can go online to search for more Gadgets.

Bring the Desktop Gadget Gallery back to the screen

On the bottom right side of the window is a link you can use to get more Gadgets (See Figure 8-7).

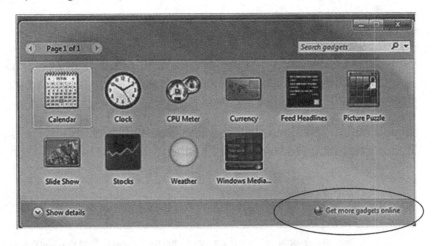

Figure 8-7

Click on the "Get more gadgets online" link

The first thing that you will notice is that you can add other themes and backgrounds from here as well as other Gadgets. You will probably want to click on the Themes tab and check out the new themes.

You will also want to click on the Desktop backgrounds tab and check out the new background images that are available.

This lesson, however, is on the Desktop gadgets and we need to look at them.

Click on the Desktop gadgets tab

There are several gadgets that you can download located on this screen. If you are not real excited yet, there are more gadgets to look at. Toward the top of the gadgets section there is a link that says "Find more desktop gadgets" (See Figure 8-8).

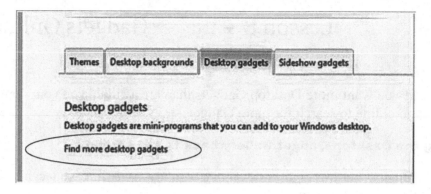

Figure 8-8

Click on the find more desktop gadgets link

Now this is more like it. At the writing of this book there are over 3,000 gadgets available. They are sorted by category and are just sitting here waiting for you to browse through the entire section.

If you find a download that you like, you will probably click on the download button. You may see something like the warning in Figure 8-9 on your screen.

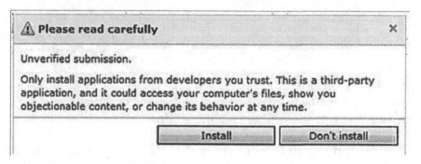

Figure 8-9

This warning means that Microsoft did not make this gadget and will not be responsible for the content. It probably works great and there will be no problems, but there are no guarantees. You have to decide if you want to install this gadget. The rest of the lesson will show you how to download and install a gadget, if you decide to do it. Since these are third party software applications and there are copyright issues, there will not be any screen shots or images used that will let you know which gadget I am downloading. I will try to talk you through this.

If you want to install the gadget click the Install button

There may be a security warning that comes to the screen. It will look similar to Figure 8-10.

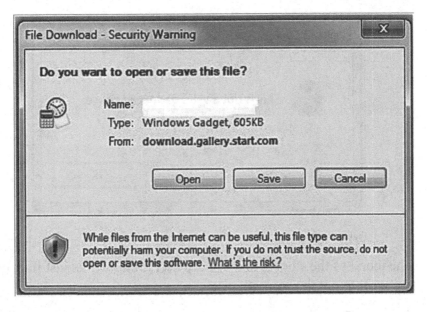

Figure 8-10

The easiest way is to click on the Open button, so that is what I am going to do.

As the file starts opening, Windows will give me one last chance to change my mind. The last security warning will come to the screen.

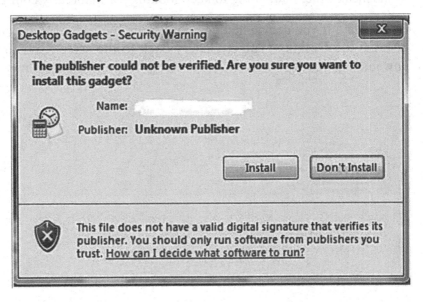

Figure 8-11

I am going to take a chance and install the gadget, so I am going to click the Install button.

Normally the gadget will install on your computer and you can start playing with your new toy, but that is not always the case. When I clicked the Install button on this gadget I received this message:

Figure 8-12

The moral of the story is that usually everything works just fine, but things don't always turn out the way you plan.

A word of caution: Don't forget some of the things that we talked about when we talked about viruses. Some people enjoy seeing if they can make your life miserable by messing around with your computer. Be careful when you download things off of the internet.

In the mean time, I am going to keep looking for that really cool gadget.

By the way, if you decide that you don't want the gadget after you have installed it, bring the Desktop Gadget Gallery back to the screen and right-click on the unwanted gadget and then choose uninstall from the shortcut menu.

Part Two

A Quick Look at some
Of the New Programs

Chapter Nine   Tablet PC

Tablet PC is a program designed to recognize handwriting and translate it to characters that can be inserted into a document. Tablet PC is designed to work with a Tablet Pen. You are more than likely familiar with these in a slightly different version. It is similar to signing the pad when you swipe your credit card at the store. A Tablet Pen will cost somewhere in the neighborhood of $100.00 up to $200.00. If you do not wish to purchase a Tablet Pen, you can also write using the mouse. Using the mouse is more difficult than using the pen.

Lesson 9 – 1   Handwriting Recognition

The Tablet PC program is located in the Accessories folder under All Programs. This is shown in Figure 9-1.

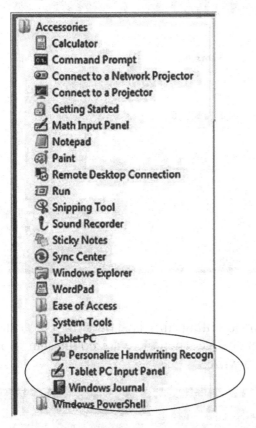

Figure 9-1

As you can see, there are three programs in the Tablet PC folder. The first program will help Windows recognize your personal handwriting. This is where you can provide samples of your handwriting that will help Windows know what you are trying to write. If you have ever used the signature pad at the store you will know that your handwriting looks very different with a Tablet Pen compared to a paper and pencil.

When you click on the Personalize Handwriting Recognition program you will get to decide how you want to teach Windows to recognize your handwriting. If there are specific characters or words that are not being recognized correctly, you can target those specific errors. If you are not sure if there are specific characters you can choose to teach the recognizer your handwriting.

If you decide to start at the beginning and teach the recognizer your style of handwriting, I would start with the numbers and letters. There are nine screens for you to provide a sample of your writing to the recognizer. The first screen looks like Figure 9-2.

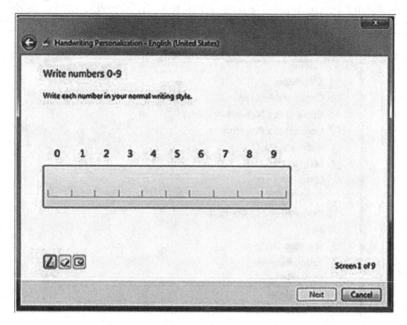

Figure 9-2

I normally have decent handwriting, this is shown to you so that you can see how your handwriting can change when there are no lines and you are watching the computer screen as you write.

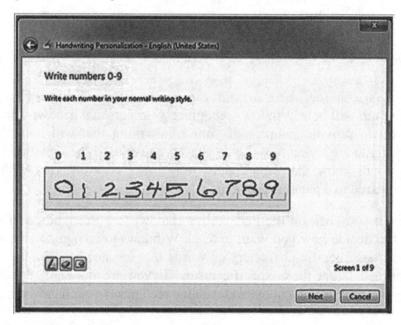

Figure 9-3

When you are finished with all nine screens, and you will probably be surprised at how bad your handwriting was, you might be asked if you would like to send a copy of your handwriting to Microsoft to help them improve the accuracy of the handwriting recognizer in future versions of Windows.

The next screen will allow you to provide more samples or update the recognizer and exit. If your handwriting looks anything like mine did, you may want to provide more samples.

Lesson 9 – 2   Tablet PC Input Panel

After you have finished and have updated your handwriting samples, you will probably want to play with The Tablet PC Input Panel. This is the second choice down in the Tablet PC folder. The panel is shown in Figure 9-4.

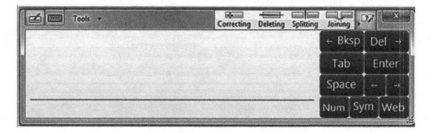

Figure 9-4

Again you can use your pen (or mouse) to write inside the yellow area and then insert the handwritten text into a word processor such as Word or WordPad. Figure 9-5 shows a sample of the handwritten text after the recognizer has checked it out and hopefully come up with the correct words. It is much better after you go through the process of providing the samples.

Figure 9-5

If you have a word processor open, even the Sticky Pad program will work, you can click on the Insert button to put the words into the document. The next line shows the "dear readers" inserted into this book.

Dear Readers

dear Readers

The Autocorrect feature in Word changed the first letter to a capital letter in example number one. In sample number two, I removed the Autocorrect and it shows the words as I wrote them.

If you need to correct a word, such as a misspelling, you need to tap the word with the pen. When you tap the word, the letters will separate so you can make corrections. Figure 9-6 shows the word "Dear" after it has been tapped by the pen.

Figure 9-6

If I wanted to change the word from "Dear" to "Deer", I would simply move the pen to the letter "a" and write an "e" over the top of it. The word would then change to "Deer".

What if I was writing the word Microsoft and I paused between Micro and soft? This is what the Input Panel would look like.

Figure 9-7

To connect the two words, I would use the pen and draw an arc under and between the two words. This would make the two words combine into one word.

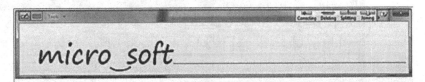

Figure 9-8

What if I wrote the word "Microsoft" and I really meant to write "Micro soft"? There is a way to fix that also. You can draw a straight line downward between the "o" in micro and the "s" in soft. This will separate the two words. I must admit that I usually have to do this several times before I can get it to work, but it will work, it just takes practice. It really is easier to use the mouse to draw a straight line than the pen. It seems to work almost every time when I use the mouse. Then again I have many years using a mouse and one day using the pen.

That should get you started using the Tablet PC Input Panel. Play with it and have some fun.

Lesson 9 – 3   Windows Journal

Windows Journal will let you save handwritten notes on your computer. Windows Journal is also designed to work with a Tablet Pen, but it will also work using the mouse if you do not have a pen. When you open Windows Journal it will look like a tablet that you can write on. Figure 9-9 shows the Windows Journal screen with a note written on it.

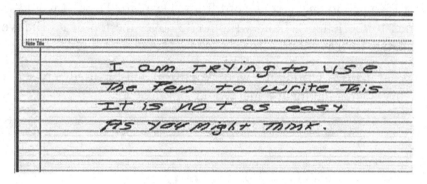

Figure 9-9

You can write anything you want and then you can save it by clicking on the "Save" icon that is on the shortcut menu or by clicking on "File" and then choosing the "Save" option from the drop down menu.

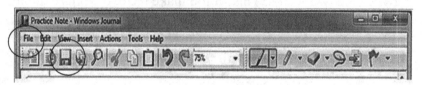

Figure 9-10

If this is the first time you have saved the journal entry, the "Save As" dialog box will come to the screen as shown in Figure 9-11.

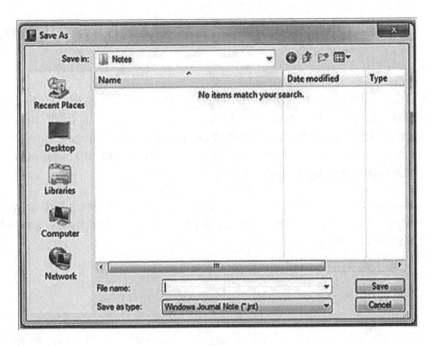

Figure 9-11

You will want to notice that the file is being saved in the "Notes" folder. All you have to do is give the file a name and click the "Save" button. You can open it later if you want to view the entry or edit it.

If I want, I can use the word processor program to write the text using the keyboard and then select the text and copy it. Then I can insert the typed text into the journal and it will show as typed text.

I can also use the Tablet PC Input Panel to write the text and insert it into the journal. It will show as typed text if I use the Input Panel.

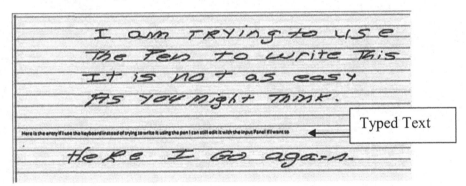

Figure 9-12

You might want to play with the journal it is kind of fun.

Chapter Ten   Math Input Panel

One of the new programs included in Windows 7 is the Math Input Panel. The Math Input Panel uses the Math recognizer that is built into Windows 7 to recognize handwritten mathematical expressions. These expressions can then be inserted into a word document or any program that supports Mathematical Markup Language (MathML).

Mathematical Markup Language was originally designed to work with Web Pages. Now many math and science software programs are now supporting MathML.

Lesson 10 – 1   Using the Math Input Panel

The Math Input Panel is designed to work with a Tablet Pen. The Math Input Panel is located in the Accessories folder under All Programs. This is shown in Figure 10-1.

Figure 10-1

When you open another program the Math Input Panel will stay on the top. The other programs will be seen behind the Math Input Panel. Figure 10-2 shows the Math Input Panel.

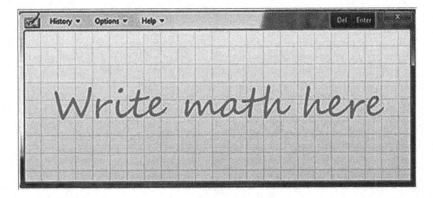

Figure 10-2

The Math Input Panel will recognize high school and college level math expressions. In my experience, the math recognizer has difficulty recognizing my expressions. As a result I have to use the correction buttons a lot of the time. Figure 10-3 shows the areas of the Math Input Panel.

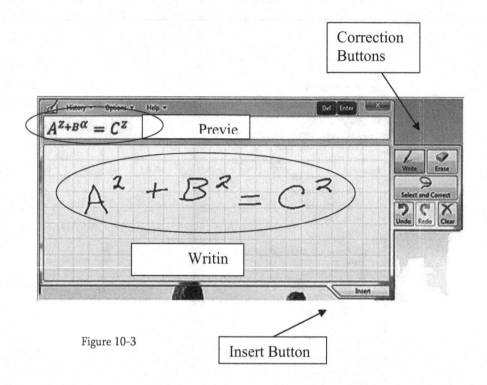

Figure 10-3

As you can see, my handwriting with the pen is not great, but I didn't think that it was so bad the preview area could not understand it. To fix the problem I need to click on the "Select and Correct" button. Then I need to draw a square around the area or symbol that needs correcting. In this case it would be the number 2 after the capitol A. Figure 10-4 shows this.

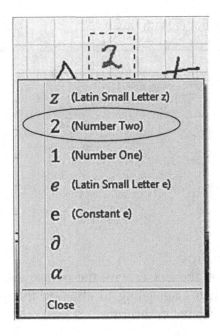

Figure 10-4

To correct this, I need to click on the number 2. This fixes the first part of the misinterpreted equation. I am not sure how it came up with "B" x "a" instead of B squared, but we can fix it.

I perform the same process of selecting the number 2 after the capital B and change it to the number 2. However the number 2 was not one of my choices.

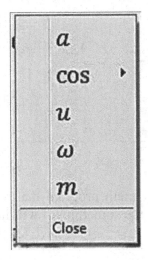

Figure 10-5

I can move my mouse over the selected area and then click on rewrite. The area inside the square will disappear and be ready for me to try again.

As you can see, it didn't get any better with this try.

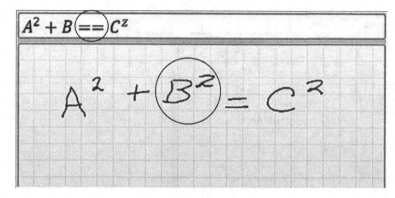

Figure 10-6

This time I am going to click on the Eraser with the mouse and then click the mouse on the problem area. Then I am going to click on the ink pen icon to write and try it again. It went much better this time.

Now we must correct the last part.

I selected the area and this time the number 2 was one of my choices. Now the formula is correct.

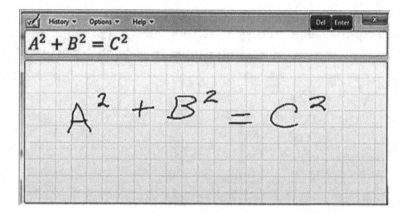

Figure 10-7

The formula can now be inserted into my math program that needs it. Just so you can see that it will insert the equation into something, I will insert it into this document.

$$A^2 + B^2 = C^2$$

It would look like I am giving this program a bad review, and at first it seemed like that would be the thing to do. However, with a little practice I can write almost any formula in one try. Windows will learn your style of writing and you also will be able to use the Math Input Panel as well as anyone else.

I am not sure how the average user will use this feature, but I have been known to be way off before, and this might be just what you need if you have a math or science program that supports this.

Chapter Eleven   Speech Recognition

Speech recognition is something that just about everyone can use and understand. You can actually use your voice to control your computer. You can say commands and your computer will respond to them. We are finally getting into the Star Trek type stuff. It won't be long until you may decide that a keyboard and mouse are so outdated you don't even want to use them anymore.

Speech Recognition was first introduced with Windows Vista and I thought it was wonderful. I used it often and had it start automatically every time I powered my computer on.

Before we get started on this chapter, you will need to have a microphone connected to your computer. If you have a laptop, the microphone is probably built-in to the computer. If you have a desktop, you will probably need to purchase a microphone.

Lesson 11 – 1   The Microphone

Before we jump in and get started, you will need to set up your computer for Windows Speech Recognition. There are three basic steps for setting up your computer:

> Set up your microphone
> Learn how to talk to your computer
> Teach your computer how to understand your speech

How quickly the computer can recognize what you are saying is going to depend on the quality of the microphone you are using. Purchasing a cheap microphone will mean that you will spend more time trying to get the computer to recognize the words you are saying. Think of it this way; I am sure you tried to talk to someone who has trouble vocalizing their thoughts and putting word together in a way that you can understand them. Perhaps their words slur together and you have to concentrate really hard just to understand them. This is the kind of trouble your computer will have if you go out a get a low quality microphone.

You should get a good quality headset microphone or at least a good quality desktop microphone. I have tried the lesser quality microphones and you will get it to work, but it is harder. The idea of the speech recognition is to make your life easier and have you rely less on the keyboard and mouse.

The first thing we need to do is set up the microphone.

Make sure your microphone is connected to your computer

If you have an external microphone is should be plugged into the red jack on the sound card.

The place where you set up the microphone is located in the Control Panel under the Speech Recognition choice.

Open the Control and make sure the All Control Panel Items is showing

Click on Speech Recognition

The following window will open and be displayed on your screen.

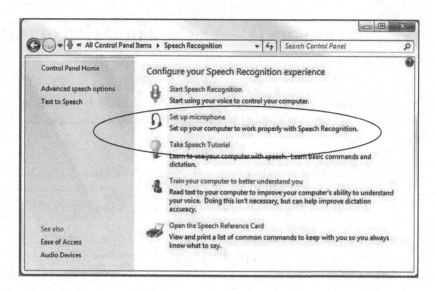

Figure 11-1

Click on Setup Microphone

The next screen will want to know which type of microphone you are using. Figure 11-2 shows this screen.

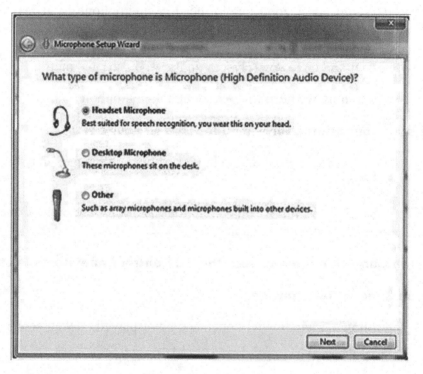

Figure 11-2

Make sure the correct choice is selected and then click the Next button

The next screen will give you some advice on how to position the microphone. Figure 11-3 shows this screen.

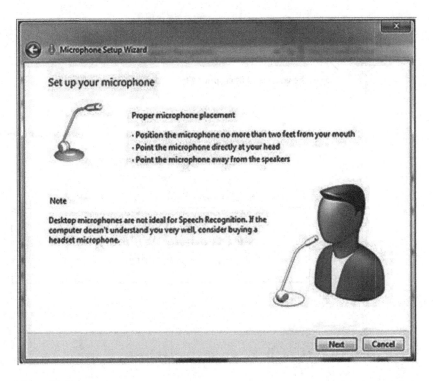

Figure 11-3

If you chose the desktop microphone you will also get a warning letting you know that a headset microphone is better for what you are trying to accomplish.

Read the directions and then click the Next button

In the next screen, you will need to read the sentence that is on the screen. This will make sure that Windows can "see" the microphone and "hear" your voice. If the results are successful, you can proceed to the next and final step. If the results are not successful, you will have to repeat this step. Figure 11-4 shows the voice test screen.

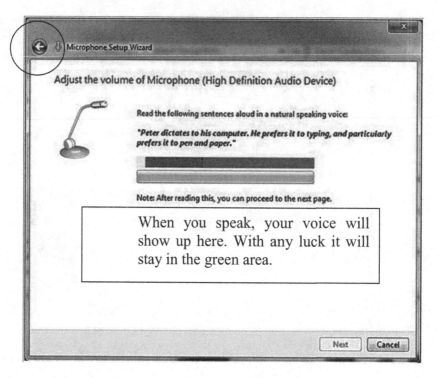

Figure 11-4

If you have to repeat this section, the Back button is circled in the figure, so you can find it.

After you have finished reading the sentence click the Next button

The last screen will tell you that the microphone is setup and ready to be used.

Click the Finish button

Figure 11-5

Setting up the microphone is pretty easy and if you use a decent microphone and speak clearly, you should not have any trouble.

Now that this is done, let's move on and see how to talk to your computer.

Lesson 11 – 2   Training your Computer

Windows comes with a speech training tutorial to help you learn how to talk to your computer. With this you will learn the commands that are used for speech recognition. In this lesson you will go through the tutorial and then you will train your computer to better understand your voice.

Open the Control Panel and then click on Speech Recognition

Click on Take Speech Tutorial

The tutorial will take several minutes. While you are taking the tutorial Windows is learning how to recognize your voice and you will be learning how to communicate with Windows by using your voice. There are certain phrases that Windows understands and you will learn these during the tutorial.

Go completely through the tutorial

Actually you will learn a lot more if you go through it twice. After you are finished, come back to the lesson.

Make sure the Control Panel is open and Speech Recognition is selected

Click "Train your computer to better understand your voice"

During this part of the lesson you will be reading text to your computer. In case you did not notice, the Control Panel says that this part is not necessary, but I beg to differ. If you are going to use speech recognition the more Windows can have access to your voice in a controlled environment the easier it will be.

Go through the training exercise and when you have gone through it once click the More Training button

I am telling you to do this because as you go through the reading part, you are also going to understand more about speech recognition. If you listen to what you are reading, you will learn a lot.

When you are finished, go back to the Control Panel and the Speech Recognition choices and open the Speech Reference Card

This is shown in Figure 11-6.

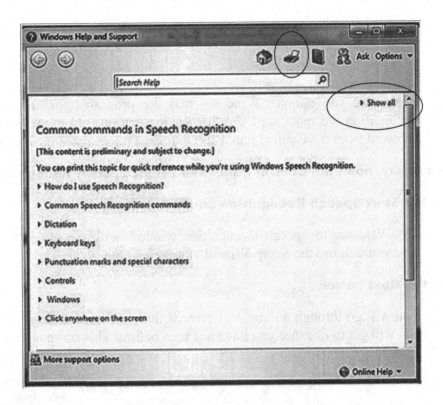

Figure 11-6

First click the Show All command and then click the Print option

This will give you a quick reference to the available commands. Keep this to use as necessary.

Wow! It didn't seem like it was going to be that long of a lesson, did it?

Lesson 11 – 3　　Starting Speech Recognition

Before we can operate Windows and the programs included using Speech Recognition we must start the Speech Recognition program. This program is accessed from the Control Panel and the Speech Recognition window.

If necessary, open the Control Panel and select Speech Recognition

Click the Start Speech Recognition command

The Welcome to Speech Recognition window will pop onto the screen. From here you will use the Setup Wizard to complete this setup.

Click the Next button

You will go through a couple of screens that you went through earlier and then you will get to one that you have not seen before. This screen is shown in Figure 11-7.

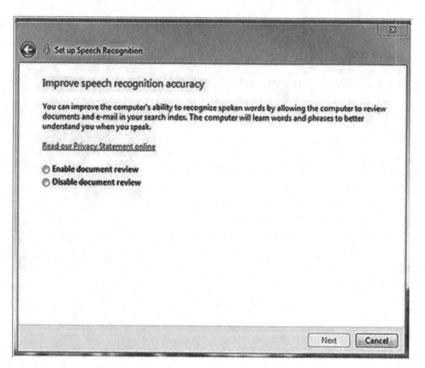

Figure 11-7

This screen will allow Windows to search through your E-mails and documents to try to help it recognize how words and sentences go together. I have not had the time to decide if this is a good thing or a bad thing. If you believe in conspiracies and Big Brother watching over you to gather information about you, you might want to disable this, just to make you feel better. If not, then go ahead and enable it. The choice is yours.

Make your choice and then click the next button

The next choice you get to make is how the Speech Recognition is turned back on if you tell the computer to Stop Listening. The choices are to tell the computer to Start Listening or manually clicking the microphone icon. Since this is about telling the computer what to do, I am going to choose the voice activation mode.

Click on the choice you prefer and then click the next button

The next screen will allow you to print the Speech Reference Card. We just did this in the last lesson, so we won't need to print it again.

Click the Next button

The next choice will allow you to start the Speech Recognition program every time you start the computer. If you plan on using it, I would say go ahead and run it at startup. If you decide not to have it start automatically, make sure the checkbox is not checked.

Click the Next button

The next screen will ask you if you want to take the tutorial.

If you want to go through the tutorial again, go ahead and click the Start tutorial button otherwise click the skip tutorial button

The Speech Recognition is now started. You can see this by looking at the top of your screen. It should look like Figure 11-8.

Figure 11-8

As you can see, mine is sleeping and to use it I will need to tell it to Start Listening. That is all there is to getting it started. Now let's see about using it.

Lesson 11 – 4   Controlling Windows

The first thing we will want to do is wake up the microphone.

If you chose the manual method, you will need to click on the microphone icon. If you chose the voice activation choice, speak into the microphone and say "Start Listening".

Either method should wake the program up. Let's see what we can do.

Say Start **into the microphone**

The Start menu should now show on the screen. If it does not open try again by speaking directly into the microphone in a clear voice. Remember when you did the tutorial, it said to try to speak as a person giving a news broadcast or as a radio spokesperson would speak.

Say Open Sticky Notes **into the microphone**

The sticky notes program should open and be in the upper corner of the screen. We have not discussed this program before. It is just like the sticky post it notes you use around the house or work. You can type a note on them and leave them on the computer screen. If you want you can use the Tablet PC and hand write the notes and insert them onto the sticky notes.

Say Close Sticky Notes **into the microphone**

The sticky notes program should close. Any program can be opened and closed this method. This is much easier than using the mouse. As a matter of fact I am using the Speech Recognition program to write this paragraph in the book. It really is easier!

Lesson 11 – 5   Working with Folders

In chapter Five we learned how to create new folders and move files into them. In this lesson we will also make a new folder and move a file into it. The difference is that in this lesson we will use Speech Recognition to do this.

Using what you have already learned about Speech Recognition open the Documents Library

When you are at the Documents Library, we will see if Windows can create a new folder using Speech Recognition.

Speak into the microphone and say New Folder

A new folder will appear on the screen and it will be highlighted. Now let's see if we can give it a new name. This should test Windows.

There are a couple of ways that you can do this. Hopefully the first way will work and we won't have to do anything silly like telling the computer to right-click on the new folder and then choose Rename from the shortcut menu (by the way, this will work if you ever need to use it).

The name of the folder should already be highlighted so all you will have to do is speak the name that you want to give the folder.

Speak into the microphone and say My letters

There are two possibilities here. The first is that Windows will change the name of the Folder to "My letters" and all will be well. The second choice, which is more probable, is that you will get to decide exactly what you want Windows to put as the title. You will probably get a small window that has something in it like the choices shown in Figure 11-9.

Figure 11-9

If you get this, you will have to say the number of your choice and then say OK. In our case, we want the first choice.

Speak into the microphone and say One

Speak into the microphone and say OK

Your new folder should now have "My letters" as its name.

Let's open the folder and see what is inside of it. If there are any files, that is where they will be located.

Speak into the microphone and say Open

The folder will open and show you its contents. Guess what? There are no contents yet. We have not put any files into the folder. Let's see what we can do about that.

Before we can move any files to the folder we will need to back up to the previous screen. We can do this by saying the command "Back".

Speak into the microphone and say Back

This will cause the computer to go back one screen. When we are back in our documents library, we can open our Windows 7 folder and copy one of our documents to the "My letters" folder.

This will require a little thinking on our part. First you have to open the Windows 7 folder and then click on a file. Then you will need to copy the file. After you have copied the file, you will need to go back one screen and then open the "My letters" folder. After we get the folder open we will need to paste the file.

Copy the Agenda file to the "My letters" folder

Try this on your own before you read any of the instructions that are printed

Speak into the microphone and say Click on Windows 7

Speak into the microphone and say Open

Speak into the microphone and say Click on Agenda

Speak into the microphone and say Copy

Speak into the microphone and say Back

Speak into the microphone and say Click on My letters

Speak into the microphone and say Open

Speak into the microphone and say Paste

If you have trouble getting the computer to click on the "My letters" file try having it click on one of the other folders and then tell the computer to press the Up or Down key until you get to the "My letters" folder.

After you get use to speaking the correct commands and the computer learns your voice, this will not seem as difficult as it seems right now.

Lesson 11 – 6   Dictating Text

Before we can dictate text we need to have a program like WordPad open.

Using the microphone open the WordPad program

WordPad, in case you have forgotten, is located in the Accessories folder which is under all programs which you get to from the Start button.

The proper sequence of voice commands would be:

> Start
>
> All Programs
>
> Accessories
>
> Open WordPad

With WordPad open all we have to do is start saying whatever we want the computer to type. Let's have the computer type the following school report for us:

Job Report

Every day after school I go to work at my job. I work at the local newspaper handing out papers to the paper carriers. I also have to clean up the office area at night before I go home. The people who work in the office can really be messy.

Not only that but down in the press room after everyone has gone home, the creepy bugs come out and chase me through the newspaper presses. You learn how to move quickly if you want to come out in one piece.

As you can tell, this is not really true, but it was fun dictating.

Give this some thought before you start speaking into the microphone

We will want the title to be center aligned and there will be three paragraphs.

When you are finished, save the file as Job Report

I hope you do not have to look at the next page, but if you need to, some of the common commands are shown.

To center the title text, first select the text and then say "Center".

To start a new paragraph, say "New paragraph".

To save the document, say "Save".

At the Save As dialog box say the file name "Job Report".

To finish saving the document say "Save".

To close the WordPad program say "Close".

With a little practice, I think you will love using the microphone.